Social Sciences and Farming Systems Research

About the Book and Editors

The range of problems faced in Third World agricultural development has taxed the capacity of research and extension methodologies based on developed country models. There remains only a partially correct fund of common knowledge regarding the reluctance of "traditional" farmers to change technologies. It is now widely accepted that the problems of technology adoption are closely tied to both the specificity of conditions necessary to use improved technologies and to the difficulties of recreating these conditions on real farms.

A major goal of Farming Systems Research and Extension is the integration of social and biological disciplines for agricultural development work in interdisciplinary research. This study reports on current FSR projects, highlighting methodological and theoretical advances in the applications of social science research in agricultural improvements. The contributions to this volume demonstrate how social science research has been applied to identify the causes of complex problems in technology change through a mixture of economic, social, and structural investigations. By providing examples, this research demonstrates how these problems can be identified and resolved within the framework of ongoing projects.

Jeffrey R. Jones is an assistant professor in the International Development Program at Clark University, Worcester, Massachusetts. **Ben J. Wallace** is chairman of the Department of Anthropology at Southern Methodist University, Dallas, Texas.

Social Sciences and Farming Systems Research

Methodological Perspectives on Agricultural Development

edited by Jeffrey R. Jones
and Ben J. Wallace

Westview Press / Boulder and London

Westview Special Studies in Agriculture Science and Policy

This Westview softcover edition was manufactured on our own premises using equipment and methods that allow us to keep even specialized books in stock. It is printed on acid-free paper and bound in softcovers that carry the highest rating of the National Association of State Textbook Administrators, in consultation with the Association of American Publishers and the Book Manufacturers' Institute.

Published in 1986 in the United States of America by Westview Press, Inc.; Frederick A. Praeger, Publisher; 5500 Central Avenue, Boulder, Colorado 80301

Library of Congress Cataloging-in-Publication Data
Main entry under title:
Social sciences and farming systems research.
1. Agricultural systems—Research. 2. Social sciences—Research. I. Jones, Jeffrey R., 1941–
II. Wallace, Ben J., 1937–
S494.5.S95S6 1986 338.'6 85-31519
ISBN 0-8133-7136-8

Composition for this book was provided by the editors.
This book was produced without formal editing by the publisher.

Printed and bound in the United States of America

The paper used in this publication meets the requirements of the American National Standard for Permanence of Paper for Printed Library Materials Z39.48-1984.

6 5 4 3 2 1

To Dr. Walter R. Goldschmidt
Pioneer in social science
research for agriculture

Contents

Tables

Figures

Foreword

Twelve years ago, I was involved in one of the first attempts to incorporate anthropologists into a national agricultural research institute in other than a communications role. Although as an agricultural economist, I had worked in such institutes myself, and had had some contact with social scientists in communication units, I had never been directly associated with social anthropologists or sociologists in **research for technology development.** As Coordinator of the Socioeconomics unit of the Guatemalan Institute for Agricultural Science and Technology (ICTA), I became convinced of the value of a good anthropologist on the first trip to the field. With little more than a casual conversation as we were driving along, the anthropologist helped me see things I had never really seen before, and become aware of the relationships that I had never thought of before, but were very relevant to a person involved in technology development. Hence, my first impression of an anthropologist was as a person with eyes to help me see things I had never before looked at and interpret things I had never before really heard.

Rapid progress has been made toward incorporating social scientists into agricultural research institutes over the past twelve years. The papers used in this book were presented at a conference that was held less than two years ago, yet a great deal has happened in Farming Systems Research and Extension (FSR/E) even since then. The annual Farming Systems symposium at Kansas State University, which this year will be held for the sixth time, has moved from papers trying to define FSR and why it is necessary to have social scientists on Farming Systems teams, to how to improve the methodology and to well written and documented case studies of successful (and unsuccessful) projects. A recent conference on Gender Issues in FSR/E at the University of Florida was notable for the absence of militancy concerning the omission of women's issues in FSR/E projects and for the very strong nature of the papers regarding methods and procedures for incorporating these issues into the ongoing operation of farming systems teams in the field. Another conference held at this University two years ago paved the way for incorporating nutritional concerns into field teams. A short course will be held this year as a follow-up.

If we realize that the concepts of farming systems as incorporated in this volume are only 10 to 15 years old, and involved the integration of many disciplines, this is amazing progress. Many, many agricultural scientists from the international community (from both national and international organizations) talk of the inputs of social scientists in the technology development process as a matter of course. Rapid appraisal, on-farm research, recommendation domains, and farmer evaluation of acceptability are taken for granted. The need to tailor technology for location specific biophysical and socioeconomic conditions is well accepted.

The concerns of the farmers with quality of the product, not just quantity of a product, and the need to stay within the resource capability of farmers are well understood by more and more biological scientists. The role of the International Agricultural Research Centers (IARCs) and the Collaborative Research Support Programs (CRSPs) to support national program research in developing countries is being worked out. USAID has just completed a new strategy to guide their research funding for Africa that recognizes the interrelationships between national level farming systems research and extension and the international research responsibilities of the IARCs and the wealthier African national research organizations. The World Bank is moving more into farming systems research and extension as the need for adapted technology has become obvious in their development efforts. Technical assistance in farming systems in Latin America is being provided more and more by other Latin Americans from countries with more experience in farming systems programs. The People's Republic of China has an active farming systems organization and held an international farming systems conference in 1985.

Much of the emphasis of this book is on farming systems as it applies to the International Agricultural Research Centers rather than on its application in the context of national agricultural institutions. The difference can seem to be petty, but it can create serious confusion if not properly understood. The IARCs do not have mandates to diffuse technology as done by the extension services of the individual countries. This would be an impossible and improper task for them. Farming Systems Research (FSR), as opposed to FSR/E, is a legitimate and appropriate strategy for the IARCs to use as a means of exposing their genetic materials and management practices to a wider range of conditions than would be possible in working only on their own experiment stations and on those of host countries. IARC scientists using FSR not only are able to test their technology on farms but are able to receive the invaluable direct feedback from the farmers on whose land the testing takes place.

National institutions, on the other hand, do have the direct responsibility for not only developing technology appropriate to the conditions in their countries, but to promote the diffusion of that technology to their farmers. For national institutions, then, FSR appropriately and efficiently becomes FSR/E, Farming Systems Research and Extension, where research and extension blend into a common effort regardless of the institutional home of the program. Researchers carry out extension functions and extensionists conduct research.

Social scientists are necessary to more efficient research in both FSR and FSR/E, but they are critical to the latter. The work of the IARCs is principally determined by biophysical considerations in what might be termed **Research domains**; a primary concern is defined on the basis of ecological conditions of interest or crop species, but only to a limited extent are socioeconomic factors taken into consideration. **Recommendation domains**, the basis for making recommendations to specifically identified farmers in FSR/E, must be characterized based on both biophysical and socioeconomic factors in order for dissemination to be efficient. **Diffusion domains**, those informal, interpersonal information networks so important to the diffusion of technology in developing countries, must be defined by social scientists so that the biophysical scientists can more efficiently locate their on-farm research in FSR/E to speed the diffusion of acceptable technologies to farmers through the recommendation domain.

Even though the term Farming Systems has spread rapidly throughout the world it is easily and often misunderstood, and connotes different things to people with dissimilar backgrounds and training. Because of this, the term itself may disappear through time. In some areas the new terms being used are Adaptive Research or Adaptive On-Farm Research. However, the impact of the farming systems concepts as described above and in this book will be felt for a long time into the future. The idea that it is necessary to understand the conditions of the farmers for whom technology is being developed in order for it to be efficiently and rapidly adopted is well accepted, as is, for example, concern about the roles women play in agricultural production and in decisions about the adoption of technology. The integration of the social sciences into the process of technology development in agriculture is also an awareness that will last.

In the present environment, this is an exciting book. It shows how socially oriented disciplines have moved through the difficult times of conceptualization and solved many of the early problems associated with the incorporation of the social sciences into the processes of biological research for development. Some of these case studies are amusing. All will serve as food for thought as more institutions take the steps required for the integration of disciplines to foster more effective research and extension programs worldwide.

Peter E. Hildebrand

Acknowledgments

The publication of this book has required the cooperation of many people along the way. From its very conception, the project of bringing together experiences from ongoing Farming Systems Research projects has faced a problem of communication due to the dispersal of the participants. Dr. William Partridge and Lynne Goldstein were instrumental in the initial presentation of the symposium on Social Science participation in Farming Systems Research at the 83rd Annual Meeting of the American Anthropological Association.

Ben Wallace has done an admirable job not only as editor but as a "point man" throughout the process of organizing the conference and preparing the manuscript. He deserves credit for expediting countless activities that could never have otherwise been accomplished because of the vagaries of international mails and telecommunications.

At CATIE, Marcelino Avila, Carlos Burgos, Luis Navarro, Miguel Holle and Bob Hart were participants and at the same time teachers in my initiation to Farming Systems Research techniques and theoretical perspectives during various projects. The excitement of working within such a dynamic theoretical environment is a rare and treasured experience.

Carlos Burgos, who participated in the establishment of all the Farming Systems Research Projects at CATIE, provided an historical perspective during the preparation of the book, and was very generous with his time in reviewing manuscript drafts during their preparation. Mario Gutierrez, forestry editor at CATIE, provided valuable editorial assistance in the preparation of the manuscript.

Editors at Westview Press displayed a remarkable patience as the manuscript preparation dragged through a number of international journeys and consultancies and breathed a heartfelt sigh of relief along with Ben and myself when customs agents around the world relented and permitted the trusty Kaypro word processor to pass unhindered and unscathed into their respective realms.

Although many of the papers presented here were carried out in conjunction with national and international research and development agencies, the views presented do not necessarily reflect the official positions of these agencies but reflect the views of the respective authors. Likewise, the views presented by the editors of this volume are their responsibility, and should not be taken to be the views of the people who have so generously helped in different aspects of the book.

Finally, my wife, Flor, and daughter Michelle have patiently endured and supported the preparation of the book. They have been the joy and inspiration that saw the project through to its end.

Jeffrey R. Jones

1

Social Science in Farming Systems Research

Jeffrey R. Jones
and Ben J. Wallace

In the course of the last few decades, world agricultural productivity has shown a marked technical change. The use of agricultural chemicals, machinery and improved seed have been diffused to the most remote outposts of agricultural production through the combined action of national and international development agencies, agricultural research centers and commercial interests. Today's farmers in developed as well as in many less agriculturally developed areas have become adept in the management of new techniques and inputs and in many cases are actively involved in informal experimentation processes to determine optimal combinations for their own agricultural, economic and social conditions (Johnson 1972). The rejection of "improved" technologies by these farmers is often based on a very correct perception that their own "traditional" technologies are superior for their purposes. The development of the FSR approach to agricultural development is a direct response to the recognition not only of the technical expertise of farmers, but also the increasing awareness and respect on the part of development technicians for locally developed production systems and the complexity of their adaptations (Harwood 1979). The fielding of interdisciplinary Farming Systems Research (FSR) teams for research and implementation activities is an attempt to ensure the identification and development of improved agricultural strategies which address the full range of constraints faced by farmers.

As part of its process of maturation, FSR has come to encompass the whole range of agricultural activities. Although first developed for use with annual crops, FSR is applied as well to animal and forestry production, which are also important aspects of small farm management (Gilbert, Norman and Winch 1980; Raintree 1984).

Social scientists have taken on an increasingly important role in the implementation of agricultural development activities. In the tradition of agricultural development, social science had been largely restricted to the participation of agricultural economists. With the recognition that farmers must respond to a broad range of social and economic conditions in the management of their agricultural activities FSR teams include and in some cases are directed by anthropologists and agricultural economists. Social scientists have entered into multiple roles including research, implementation and management, in support of overall of FSR project objectives.

This book grew out of a symposium held in November of 1984, which was convened to discuss the activities of social scientists in FSR. The participants in the symposium worked on FSR projects for regional and international agencies in the direct implementation of agricultural development programs. The primary objective of this book is to make available to a wider audience the results of this conference, and to stimulate the discussion and analysis of the role of social scientists in FSR.

A second objective of this collection of papers is to open a social science discussion of the potential theoretical contribution of FSR. The breadth and volume of FSR project experiences make them a "real world laboratory" to compare individual and social responses to change and development, and at the same time offer a possibility for operationalizing and implementing social science concepts. The changes in world agriculture during the past few decades have radically altered the conditions of third world farmers and an evaluation of experiences in the implementation of development projects will necessarily lead to the re-evaluation of, and improvements in, the concepts and assumptions which underlie models of development.

FSR in International Agricultural Development

Since the 1950's, the implementation of international agricultural development projects was accompanied by a body of critical literature documenting development failures and unwanted side-effects of the development efforts. A collection of papers by Spicer (1952) highlighted social and cultural conflicts in the introduction of new techniques, such as the resistance to new corn varieties because of taste and mechanical properties. Foster's review of technological change identified social conflicts and technical shortcomings in the introduction of new technologies (1962), and Epstein (1973) documented the exacerbation of inequalities in income distribution as a consequence of the use of these technologies[2]. Griffin (1972) documented an even more disturbing "development" outcome, where not only was income inequality emphasized, but total agricultural production in fact decreased with the introduction of technical improvements. These are only a few of many similar critiques which pointed to an alarmingly persistent pattern of unwanted results in agricultural development efforts.

One strategy which had provided the most promising results in agricultural development were the genetic improvement techniques of the "Green Revolution". The possibility of producing genetically improved crops to poor farmers is widely regarded as the best possibility for increasing agricultural yields and improving living standards. International Agricultural Research Centers such as IRRI (the International Rice Research Institute at Los Banos, in the Philippines), CIP (the International Potato Center, at La Molina, Peru), CIMMYT (the International Center for Corn and Wheat Improvement in Mexico), etc., were established in different climatic zones to investigate the possibilities of improving major food crops, such as corn, wheat, rice, beans, cassava, potato, etc. Nevertheless, the techniques developed by these centers were also the object of the criticism that agricultural solutions created social and economic problems which tended to offset their overall benefit.

New methodologies for preliminary research were needed to identify target populations and their general conditions as a means to orient the research of the international centers. This research would have to identify not only the agricultural problems of the intended target populations, but also try to foresee patterns of use and potential problems for new technology. Yield improving research tends to focus on optimal production conditions, despite the fact that the majority of farmers do not work under optimal conditions; since farmers who do work under near optimal conditions are likely to be wealthier than others, yield improvement research for optimal conditions would tend to exaggerate wealth differences. By identifying constraints specific to target populations defined in terms of socioeconomic or agroeconomic conditions, research efforts would have higher probabilities of developing technologies to improve the conditions of those populations.

The problem bequeathed to FSR was, simply stated, to construct an integrated picture of peasant production, formulate strategies for identifying problems for specific farming populations, and then to propose and develop solutions. The strategy of integrating the contributions of biological and social scientists at every stage of the research process is designed to ensure that problems were correctly perceived both in a social and technical sense. While this integration had been discussed and tried previously, the FSR approach made exceptional efforts to guarantee that this integration was functional and not merely cosmetic.

Farming Systems Methodology

Several different approaches to FSR were developed nearly simultaneously at different national, regional and international agricultural research centers and different variants are recognized such as FSR/E (Extension) and FSR/D (Research and Development) depending on the problem focus. The Cropping Systems approach was developed at IRRI as a response to the need perceived at that institution for combined crop rather than single crop production strategies, given the production conditions of South East Asian farmers (Zandstra 1982); Shaner **et. al.** (1982a) found Cropping Systems Research to be very similar to FSR when a broad set of research selection criteria had been used to establish the need for the cropping systems focus. One of the first agricultural centers that applied FSRM was ICTA (Instituto de Ciencia y Tecnologia Agricola) in Guatemala (Hildebrand 1979). At the same time, other centers such as the Tropical Agricultural Research and Training Center (CATIE) (Moreno 1977, Moreno and Saunders 1978, Navarro 1979, Hart 1980), CIMMYT (Byerlee **et. al.** 1980, Winkelmann and Moscardi 1982) and CIAT (the International Center for Tropical Agriculture in Cali, Colombia) were working along parallel lines and applying the results in their research. The widespread use of FSR in CGIAR centers (Consultative Group for International Agricultural Research) led to reviews of these activities, one in 1978 by a CGIAR appointed Technical Advisory Committee (TAC 1978), and another by a University based program in 1980 (Gilbert, Norman and Winch 1980). A more encyclopedic review of FSR methodologies was later produced by Shaner, Philipp and Schmehl in 1982a (Shaner **et. al.** 1982a). The process of consolidation of FSR in its definitive form is far from complete; it can be noted that even in this volume, the team from CIP still finds itself uncomfortable with the FSR label, despite the obvious similarities between the CIP approach and the more generally recognized FSR methodology. The model for FSR discussed here was developed at CATIE and is presented recognizing that it is one of several models developed at different institutions which bear a general similarity. This model is presented because of its detail in the crucial technology testing phase of FSR, and has been discussed in even more detail in the different CATIE publications cited above.

The most significant characteristic of the different FSR approaches is their attempt to take biological experimentation to farmers' fields fairly early in the research process, and to build farmer feedback into evaluations at various points. As technological alternatives are developed, be they new crop varieties, the use of new inputs, or new management techniques, they are taken off the research station to agricultural zones where they are likely to be applied. The observation of the new techniques under local soil, rainfall, and pest conditions may provide an early warning of problems that could have been overlooked at the experiment station. At the same time, an attempt is made to achieve a maximum participation of farmers on whose land trials are planted. In effect, the process requires the identification of agricultural "experts" who understand their local conditions, who have a motivation to discover new production techniques and who are capable of communicating to the researchers their perceptions of the progress of the experiment and its potential applicability to their farm situation.

The need for farmer "experts" in the research process derives from the problems faced in real-life farm management (FSR is seen by some as having clear links to the academic study of farm management (Gilbert, Norman and Winch 1980). When all factors, including environmental conditions, socioeconomic conditions, and farmers' objectives in the management of their farms are taken into account, farm management strategies are extremely complicated. Although university trained experts can afford to specialize in forestry, or annual crop or animal production, the farmer must manage all the components of his farm, not only for production of each component, but to maximize the productivity of the farm as a whole in the context of the interactions between components. The final outcome of alternatives to the farm system and the corresponding interactions throughout the farm can only be known empirically.

As a response to the problem associated with the complexity of peasant farms, FSR became an interdisciplinary concern. This approach avoids the disciplinary bias which may cause the investigation to overlook factors of importance which either overlap the boundaries of disciplines, or which are not strictly biological. The composition of an FSR team is not fixed; it should include biological and social scientists, but within these broad outlines there is a range of variation (see TAC 1978 and Hildebrand 1976 for discussions of team composition).

Social Science Inputs in FSR Research

Agricultural development has slowly come to accept the

value of social science research. Where the problems of agricultural development were once conceived as being eminently biological, FSR is now seen to address "1) the interdependencies among the components under the farm household's control and 2) how these components interact with the physical, biological and socioeconomic factors not under the household's control" (Shaner **et.al.** 1982b).

The papers in the present volume were brought together under the conditions that 1) they addressed social, biological and economic issues relating to agricultural development questions and 2) they were carried out in the framework of interactive development and implementation projects. The restrictive concern here is necessary to differentiate FSR from more traditional holistic research in agricultural societies. Numerous social science investigations have looked into the interrelations of physical, biological and socioeconomic factors, but would not be considered FSR; Hill's analysis of West African cocoa growers (1963), Cancian's work on peasant social structure and economics (1965), and the entire school of "cultural ecology" (Netting 1977) all address the broad question of the relation between physical and social factors. These studies, however, were not directed toward the resolution or clarification of specific questions of importance for the implementation of new agricultural technologies.

What is of special interest in this volume is the set of investigation techniques which have been brought to bear on agricultural implementation problems. FSR requires new combinations of research strategies to acquire relevant information in a timely fashion for use within a project framework. This constraint has given rise to some innovations in field work. At the same time, it is striking that certain basic social science research strategies are reinforced by the experience in FSR, reaffirming their fundamental validity and usefulness.

The Research Process in FSR

FSR is an interactive research strategy characterized not only by the use of interdisciplinary teams but also by the integration of experiment station research with socioeconomic investigations and on-farm trials of technologies in the field. The flow of information and feedback permit a constant refinement of activities in an iterative fashion to better address the conditions of the target population.

The first stage of FSR consists of the definition of

project objectives and **work area selection.** Under ideal conditions these decisions should be based on broad policy objectives, such as improving living conditions for a given socioeconomic stratum or increasing employment. The realities of development financing and management often do not permit complete freedom in this regard due to institutional and policy objectives of host countries. This first stage does not necessarily require field investigation and may rely primarily on census data.

The next stage consists of a **characterization** of the work area (this discussion closely follows Figure 1.1). This activity is carried out by interdisciplinary teams in a short, intensive work period called a **sondeo** by Hildebrand (1979). The objective of the characterization is to identify the hierarchy of systems of the work area including regional and intersectoral linkages, and to define predominant "farming systems" (see CATIE 1982 for an example of a regional characterization including regional systems and hierarchies). The farming systems defined are spatial or temporal juxtapositions of crops, either with other crops, or animals, or with trees. The systems also include functional associations, in which products of one activity form an input for some other farm activity, e.g. a crop which is used as fodder for animals, or green manure for other plants. The function of the characterization is to define systems which will influence the primary objective of the project. Furthermore, the characterization helps to interpret census data in the definition of frequencies and distribution of the farming systems of interest.

The characterization serves as a basis for the **initial identification of problems and systems** within the work area. All the relevant elements of the agroecosystem in question must be identified in order to avoid incompatibility when the phase of technology transfer arrives; in addition, the project focus must be production systems which are widely distributed and are sufficiently homogeneous to allow for the application of a common technological improvement strategy over most of their extents. By taking these points into account, the **design of improved systems** will be directed toward farms of the target population, and will be applicable to a large enough number of farms to justify the investment in research.

Design classification depends on the degree of confidence in the proposed technology and the quality of data available for each of the components in the system. If there is an abundant experience in the field with similar agroecosystems, data will be more reliable than data which is derived only from the experiment station. When there are sufficient data regarding the applicability of the technology, it can be **validated** immediately under field conditions. The process of **validation** is

FIG. 1 FARMING SYSTEMS RESEARCH

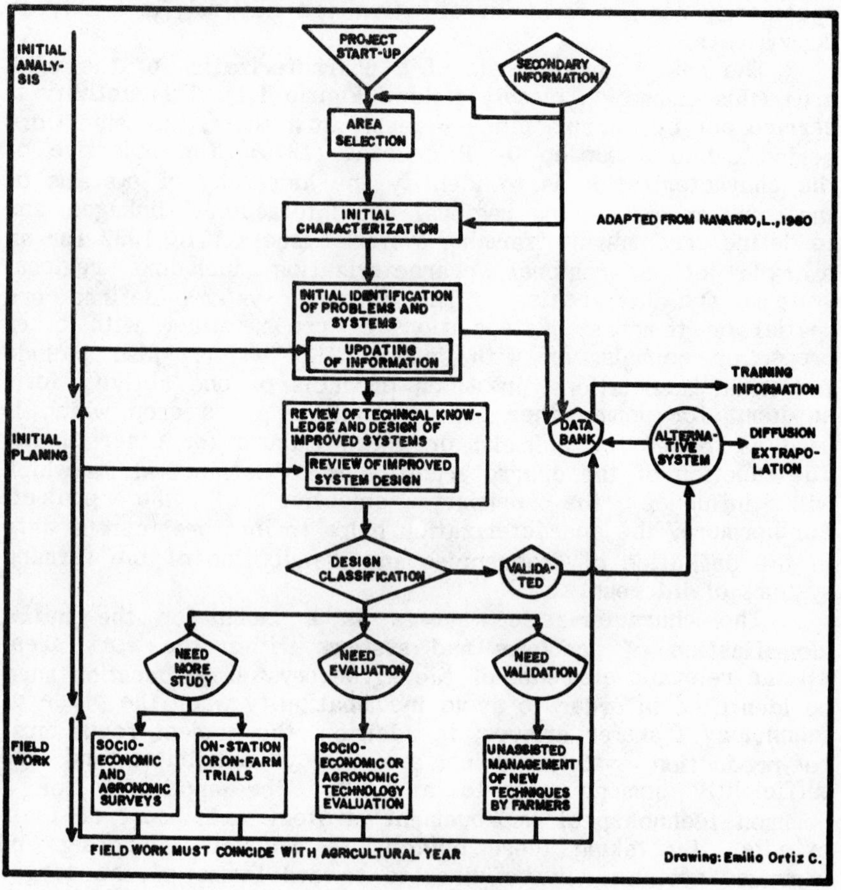

INITIAL ANALYSIS
PROJECT START-UP
SECONDARY INFORMATION
AREA SELECTION
INITIAL CHARACTERIZATION
ADAPTED FROM NAVARRO, L., 1980
INITIAL IDENTIFICATION OF PROBLEMS AND SYSTEMS
UPDATING OF INFORMATION
TRAINING INFORMATION
REVIEW OF TECHNICAL KNOWLEDGE AND DESIGN OF IMPROVED SYSTEMS
INITIAL PLANING
DATA BANK
ALTERNATIVE SYSTEM
DIFFUSION
EXTRAPOLATION
REVIEW OF IMPROVED SYSTEM DESIGN
DESIGN CLASSIFICATION
VALIDATED
NEED MORE STUDY
NEED EVALUATION
NEED VALIDATION
FIELD WORK
SOCIO-ECONOMIC AND AGRONOMIC SURVEYS
ON-STATION OR ON-FARM TRIALS
SOCIO-ECONOMIC OR AGRONOMIC TECHNOLOGY EVALUATION
UNASSISTED MANAGEMENT OF NEW TECHNIQUES BY FARMERS
FIELD WORK MUST COINCIDE WITH AGRICULTURAL YEAR
Drawing: Emilio Ortiz C.

characteristic of the FSR approach (Shaner **et.al.** 1982 discuss this procedure under the heading of "acceptability"). The improved technology is taught to farmers during an initial period covering one production period or one agricultural year, after which they are left to their own devices to manage the alternative in the form they find most appropriate. Validation itself is the observation of the use of the technological alternative under farmer management, with whatever technical advice they request. A critical point which differentiates validation from on-farm trials is that all normal production costs are assumed by the farmer in validation, so that these will form a part of the farmers' decision making in the acceptance of improved technologies. This duplicates real production conditions, except that farmers have participated in the initial training period to learn how to manage the technology. Variations in the patterns of management by farmers are studied to determine which are adaptations to special conditions encountered in the area.

If the technology is generally accepted by the farmers in the validation sample, it may be passed to the stage of **recommendation.** If this is the case, it is presented to ministries and extension agencies for wider diffusion. More commonly, the technological alternative does not pass directly to recommendation, since there are likely to be some impediments to its generalized application. For further refinement, the alternative is returned again to earlier stages in the process. It may be necessary to collect more socioeconomic information regarding farm management (see esp. Conklin's (1982) discussion of social science information necessary for technology implementation), or more biological research into different aspects of the technology, or possibly a more general reconceptualization of the alternative (note feedback in Figure 1.1).

One possible route for development is followed when the initial classification of the technological design does not permit sufficient confidence in the technical proposal to warrant field trials. In that case, it may be necessary to limit investigations to experiment station work and field observations of farming systems similar to the proposed alternative.

An intermediate path is that of technology **evaluation.** This path may be used when the information which is lacking to implement an alternative technology depends on the on-farm interaction of components. In evaluation, the project technicians actively participate in the management of the experiment, since the objective is to gather more information on system performance under specific conditions, rather than the observation of independent farmer management. Technology evaluation may require either biological or socioeconomic investigations, depending on the problems faced.

It is important to note that in the phases of evaluation and validation, there is considerable interchange between farmers and technicians. This is a two-way communication. The technicians teach farmers alternative techniques, but at the same time try to understand the local limitations which confront the farmers. The technicians also try to understand farmer preferences which shape their management strategy. This interaction is what gives FSR the potential for developing alternative technology with a minimum of resources wasted in the refinement of techniques which have little possibility for acceptance due to reasons independent of their agricultural efficiency. By understanding the farmers' management strategies, and the objectives and limitations which shape these strategies, there is a better possibility of designing technological alternatives which resolve problems encountered by farmers in the improvement of their agricultural enterprise.

In conclusion, FSR methodology can be broken down into five phases;

1. The identification of the project area and problem focus.
2. Characterization of the work area on a geographic basis, as well as on the basis of farming systems.
3. Design of alternative technologies.
4. Validation of technologies.
5. Recommendation of validated technologies.

The feedback between phases built into the FSR strategy is especially important to the FSR design. It is not a unilinear process and admits the possibility that technologies may be identified which at some point in the investigation process must be abandoned. This does not imply a failure in the design process, but rather reflects the complexity of the target systems. The need to abandon a technological alternative only becomes a failure when its shortcomings are not acknowledged until an inordinate amount of resources has been invested in its development. Another important point is that the reactions of the farmer enter into the process of design and testing through the processes of characterization, evaluation and validation. Technological problems are identified early in the project so they can be corrected and tested within the project time frame.

The use of interdisciplinary teams permits a rapid and objective evaluation of problems and the definition of project foci. While this implies a fairly large initial investment in the project, if properly executed it reduces costly errors which could result from overly narrow or biased initial evaluations of problems.

FSR is an experiment on several different levels. On a technical level, it is an attempt to develop and maintain a flow of information and a set of working relationships in an interdisciplinary team so that information is immediately available to be incorporated into planning and implementation processes. Decisions as to objectives in biological trials should be made on the basis of social science information, and areas of interest for social research should be directed toward defined areas of project interest (see Rhoades **et.al.**, and Brush in this volume). On an administrative level, it formalizes the relationship between the disciplines and prescribes a specific "group dynamic" to achieve integration within the team.

Given the complexity of the objectives of FSR, the patterns of implementation are highly variable. The process of learning about local conditions and adapting to local needs introduces a certain unpredictability in the project content, and generates a diversity among projects, although all are based on a common methodology.

Theoretical Implications of FSR

The most significant theoretical contribution of FSR is the recognition and in a sense the validation of the technical viability of traditional agricultural systems. This contribution is tied to the observation of 1) the diversity of viable agricultural strategies on both a worldwide and a regional level 2) the functional integration of social and biological aspects of agricultural production and 3) the persistence of small farm (peasant) agriculture in its generally universal role of provisioning entire societies using traditional and scientifically unstudied techniques.

The diversity of agricultural techniques has been summarized in the the works of Ruthenberg (1980) and Harwood (1979). Their contribution is not the identification of "new" systems; anthropological research has documented a wide variety of systems, from low intensity swidden farming (Conklin 1955) to highly intensive rice production (Geertz 1967), and has included household level economics (Tax 1953) and market level phenomena (Beals 1980). What is new in the writings of Ruthenberg and Harwood is the attempt to digest and synthesize the same sort of information with the objective that it be incorporated into a general strategy of agricultural development, and its **recognition as such by agricultural researchers and technicians.**

An important motivation for these developments in FSR is the realization that peasant farmers are very knowledgeable about the environment in which they work, and the crops they plant. It hardly needs to be pointed out that all major food crops were originally developed by peasant farmers. There is also evidence suggesting that valuable indigenous cropping systems exist which still remain to be tapped by agricultural scientists in their efforts to find more effective ways to improve production and human well-being. The practice of intercropping has been most extensively developed by peasant farmers, and experiment station technicians are still trying to understand and improve only a few of the known systems[3]. Similarly, "new" crop species and "new" varieties of common crops are constantly being identified for investigation by international organizations exemplified by the International Bureau for Plant Germplasm Resources (IBPGR based in Rome). The investigation of these systems requires a special "ethno-agronomic" approach to understand these crops and farming systems, their proper management and their potential for improvement.

Although unrelated to FSR in its origin, the concept of Indigenous Technical Knowledge is very closely allied to the interest of FSR. This phrase has been proposed for the investigation of existing native technical systems (Brokensha, Warren and Werner 1980), focusing on the identification of technical strategies which are adapted to particular environments, but whose management is largely an unwritten tradition. While Conklin's study of Hanunoo agriculture is the theoretical progenitor of these ideas, numerous more recent investigations of indigenous agricultural technical knowledge exist (Hatch 1976; Brokensha **et.al. ibid** has several articles and a useful bibliography).

FSR implicitly recognizes the need to understand a diversity of agricultural systems in the search for solutions to agricultural development problems. The understanding of the context of agricultural techniques and the specific problems they address leads to a more complete view of the social and economic forces in developing societies (Orlove 1977; Guess 1977; Barlett 1982), which help agricultural scientists learn new techniques from farmers, discover new crops and understand alternative problem solving strategies developed to deal with problems in different environments. These development needs coincide closely with certain social science research interests, and provide an important opportunity for further collaboration.

The interaction of cultural and biological factors has been the subject of anthropological investigations under the heading of Cultural Ecology (Netting **op.cit.**). The need to understand the motivations of farmers and constraints on agricultural technology poses questions which can be best answered from within this theoretical framework (Wallace 1984). The interaction of FSR and Cultural Ecology may well generate both technical and theoretical data which will be of mutual benefit to agricultural development and social science theory.

Certain recent works promise to link Cultural Ecology more closely to agricultural development concerns. In the past, cultural ecology has tended to focus on small scale societies, but there seems to be more interest recently in considering peasant societies and their economic and social links to broader societies (see esp. Bennett 1976). The application of FSR provides a context where this sort of information can be generated and where conclusions are likely to be tested through the implementation of new production systems (DeWalt 1985).

Perhaps a more significant contribution of FSR to social science theory, and the one which has a more far-reaching impact, is the recognition that non-mechanized, diversified peasant farming systems are a permanent feature of many societies, and make major economic and social contributions to the reproduction and maintenance of those societies. At one time it seemed to be an unquestioned truth that agricultural development must ultimately follow the North American pattern of mechanization, elimination of small farmers, and the formation of oligopolies. Nevertheless, peasant farmers have outlived the predictions of their demise in both capitalist and socialist development. The numerous failures of the industrial agricultural model in developing countries raise the question of the applicability of the large scale "industrial" approach. The persistence of peasant farming has been found to be theoretically viable in an economic sense (Chayanov 1928) and empirical data have confirmed these arguments (CIDA 1962). FSR implicitly accepts these observations and begins to seriously focus scientific research attention on the understanding and improvement of the peasant agricultural systems. While FSR initially had been promoted as a methodology for promoting social and economic equity rather than economic efficiency in development, a strong case can be made for its making major economic contributions to agricultural development (cf. Schultz' (1964) discussion of the potential of farming populations in development).

This last lesson may be one of the most important to arise from the agricultural development process. Increasing wealth disparities among certain agricultural sectors have emphasized the specificity of agricultural technology, and the different capacities of different agricultural populations. The efforts for agricultural development must be directed toward the discovery of alternative agricultural models as well as toward the adaptation of existing technology to developing country environments. A development strategy such as FSR which begins with the observation of local production techniques and makes constant "course corrections" in response to built-in feedback mechanisms promises positive development for local production systems.

Agricultural development in effect is a continuing empirical test of social and economic theories of development. These experiences must be continually evaluated to avoid repeating mistakes. In the field, these evaluations are made in a rather "rough-and-ready" fashion, with the objective of project implementation rather than logical coherence or theoretical elegance. Unfortunately, "rough-and-ready" analyses contribute principally to personal insights, rather than institutional or theoretical development. The bridging of the gap between empirical implementation insights and theory will be constructive for both the academic social sciences and the process of agricultural development.

Volume Contents

The volume is divided into four topic areas; general concepts, annual crop production, animal production and forestry. Nevertheless, due to the nature of FSR these topic areas are highly permeable. All the papers refer to general concepts and methodological orientations in FSR, and the papers within the specific production categories refer to other production areas in their attempt to understand farming systems as wholes.

The composition of teams is a fundamental problem in FSR. Rhoades, Horton and Booth (Chapter 2) document the tension which exists within research teams caused by the use of different disciplinary assumptions in the identification of problems and solutions. The resolution of the tensions within the team illustrates one of the important strengths of FSR, the ability to generate a range of working "hypotheses" and then resolve them through the interaction of team members and the execution of field activities.

Brush (Chapter 3) also deals with conflicts between researchers, but in the area of "basic" versus "applied" research. These conflicts reflect disciplinary projections of research priorities based on past experience and current capabilities. The fusion of disciplinary perspectives is an important process in basic research, and the experience of CIP, in Peru, provides a valuable model.

Chapman (Chapter 4) touches a problem at the heart of agricultural development, the design of alternative technologies. The problem of controlling costs of research is closely related to the ability to choose the right problem in the first place, before making large scale investments. A method for sifting through technical alternatives is illustrated, which considers technical-biological capabilities, economic and social data to improve the probabilities that researchers will receive no unpleasant surprises when technologies are introduced to farmers.

The acceptability of research on alternatives is improved through the validation of technology and technology transfer strategies under farm conditions. Wallace (Chapter 5) and Hansen (Chapter 8) discuss how biological researchers can be divorced from local realities by their research procedures, resulting in problems when technology is tested in the field. Planning problems in the establishment and use of demonstration plots were discovered to limit the effectiveness of the demonstration procedure and the transmission of useful information to farmers.

When technologies are transferred to farms, a wide variety of factors come into play, in addition to the biological appropriateness of the alternative. Reeves (Chapter 6) analyzes the structural constraints on marketing opportunities, and how these interact with farmer income levels to influence adoption. This analysis suggests some factors to observe in the economic evaluation of technological alternatives to guarantee broader applicability and avoid marketing constraints which may make them only partially acceptable to farmers.

Spring (Chapter 7) discusses how ethnocentric farming assumptions of male-female roles can result in overlooking a major portion of the farming population. A more realistic assessment of women's contributions to farming dictates the inclusion of extension programs specifically directed toward women farmers. In the course of this work it also became necessary to refine teaching methods for effective technology transfer.

Animal and forestry production systems present special problems for agricultural development. Relatively long production cycles, and the combination of crops and animals, or crops and trees, create complex systems whose analysis oversteps the limits of on-station research.

Jones' discussion of a Central American mixed animal-crop production project (Chapter 9) suggests how long production cycles can be analyzed within project time frames. The systematic collection of farmer perceptions in many cases provides a check on long (and short) term acceptability of technologies.

Michie (Chapter 11) and Robins (Chapter 12) document the analysis of farm-forest relationships in two project areas. Current needs and management practices of forest areas are shown to be important for both agricultural and forestry development, the understanding of which forms a basis for future research and implementation strategies.

Curry's work in an African livestock project (Chapter 10) reaffirms the importance of basic ethnographic techniques for project development. The ability to quickly gather and analyze ethnographic information through anthropological techniques emphasizes the need for a range of data gathering procedures to ensure that information is available in a usable form, and when it is needed.

Taken as a whole, these analyses present examples of social science involvement in FSR which emphasize the continuity of fundamental problems in agricultural development and illustrate how FSR techniques are being employed to address the problems. While their diversity may seem confusing if a simple answer to the problems of development methodologies is sought, the collection thereby attempts to present a realistic picture of social science participation in FSR.

Notes

1. The symposium was entitled "Farming Systems Research; the Integration of Social and Biological Approaches for Agricultural Development". It was held in Denver, Colorado, at the 83rd annual meeting of the American Anthropological Association. The original symposium participants included Walter Goldschmidt, Billie DeWalt, Timothy Finan, Sheldon Smith, and Kjell Enge, in addition to the contributors to this volume.
2. It is worth remembering the reaction to the study of social impacts of large industrial farms in the Central Valley of California (Goldschmidt 1947, republished in 1978). A major finding of that report was that agricultural mechanization and the concentration of productive resources had visibly negative social consequences. At that time it was argued that social scientists had no business doing that sort of research, and extreme political pressure was exerted to impede its publication (Taylor 1976, Goldschmidt 1978). It is curious that the criticisms of Green Revolution technology should seem to have come as such a surprise to some.
3. A revealing discussion of the "discovery" of intercropping within the context of the Plan Puebla is presented in Whyte (1985), and Gilbert, Norman and Winch (1980) note a similar situation in Africa.

Bibliography

Barlett, Peggy. 1982.
Agricultural Choice and Change. New Brunswick, New Jersey. Rutgers University Press.

Beals, Ralph L. 1975.
The Peasant Marketing System of Oaxaca, Mexico. Berkeley: University of California Press.

Bennett, John. 1976.
The Ecological Transition: Cultural Anthropology and Human Adaptation. Chicago: Aldine.

Brokensha, David, Warren and Werner. 1980.
Indigenous Technical Knowledge and Development. Lanham, Maryland: University Press of America.

Byerlee, D., S. Biggs, M. Collinson, L. Harrington, J.C. Martinez, E. Moscardi and D. Winkelmann. 1980.
Planning Technologies Appropriate to Farmers. El Batan, Mexico: CIMMYT.

Cancian, Frank. 1972.
Change and Uncertainty in a Peasant Economy. Stanford, California: Stanford University Press.

CATIE (Tropical Agriculture Training and Research Center). 1982.
Caracterizacion de Sistemas Agricolas de Hojancha, Guanacaste, Costa Rica. Turrialba, Costa Rica: CATIE.

Chayanov, A.V. 1928.
Peasant Farm Organization. Republished in D. Thorner, B. Kerblay and R.E.F. Smith (eds.) 1966. **The Theory of Peasant Economy.** Homewood, Illinois: American Economic Association.

Conklin, Harold. 1957.
Hanunoo Agriculture in the Philippines. Rome: FAO.

Conklin, Frank S. 1982.
"FSR/E as a Management Tool for Use in Third World Countries: An Historical, Institutional and Economic Perspective with Suggestions for Change". Paper presented at symposium "Review of Farming Systems Research", South Dakota State University. (Mimeo).

Comite Interamericano de Desarrollo Agricola (CIDA). 1962. **Peru - Tenencia de la Tierra y Desarrollo Socio-economico del Sector Agricola.** Washington, DC: OEA-BID-FAO-CEPAL-IICA.

DeWalt, Billie R. 1985. "Anthropology, Sociology and Farming Systems Research". **Human Organization** Vol. 44 (2); 106-114.

Epstein, T. Scarlett. 1973. **South India: Yesterday, Today and Tomorrow.** London: Macmillan.

Foster, George M. 1962. **Traditional Cultures and the Impact of Technological Change.** New York: Harper and Row.

Geertz, Clifford. 1963. **Agricultural Involution.** Berkeley, California: University of California Press.

Gilbert, E.H., D.W. Norman and F.W. Winch. 1980. **Farming Systems Research.** East Lansing, Michigan: Department of Agricultural Economics, Michigan State University.

Goldschmidt, Walter R. 1978. **As You Sow: Three Studies on the Social Consequences of Agribusiness.** Montclair, New Jersey. Allanheld, Osmun and Co.

Griffin, Keith B. 1972. **The Green Revolution: An Economic Analysis.** Geneva: United Nations Research Institute for Social Development.

Guess, George M. 1977. **The Politics of Agricultural Land Use and Development Contradictions.** Ann Arbor, Michigan: University Microfilms.

Hart, Robert D. 1980. **Agroecosistemas.** Turrialba, Costa Rica: CATIE.

Harwood, Richard R. 1979. **Small Farm Development: Understanding and Improving Farming Systems in the Humid Tropics.** Boulder, Colorado: Westview Press.

Hatch, John K. 1976. **The Corn Farmers of Motupe.** Land Tenure Center Monograph #1. Madison, Wisconsin: University of Wisconsin.

Hildebrand, Peter E. 1976. "Generating Technology for Traditional Farmers: A Multidisciplinary Methodology." Paper prepared for the conference on" Developing

Economies in Agrarian Regions". Bellagio, Italy.

———. 1979. **Incorporating the Social Sciences into Agricultural Research.** New York: ICTA (Instituto de Ciencias y Tecnologia Agropecuaria, Guatemala) and Rockefeller Foundation.

Hill, Polly. 1963. **Migrant Cocoa-Farmers of Southern Ghana.** London: Cambridge University Press.

Johnson, Allen. 1972. "Individuality and Experimentation in Traditional Agriculture". **Human Ecology** 1; 149-159.

Moreno, Raul. 1977. "Sistemas y Enfoque de Sistemas". Turrialba, Costa Rica: Departamento de Produccion Vegetal, CATIE. (mimeo).

Moreno, Raul and Joseph Saunders. 1978. "A Farming Systems Research Approach for Small Farms of Central America." Turrialba, Costa Rica: Departamento de Produccion Vegetal, CATIE. (mimeo).

Navarro, Luis A. 1979. "El Problema General de la Agricultura y la Investigacion Agricola Basada en el Enfoque de Sistemas." Turrialba, Costa Rica: Departamento de Produccion Vegetal, CATIE. (mimeo).

———. 1980. "Una metodologia para el desarrollo de tecnologia agricola apropiada para pequenos agricultores de un area especifica". Paper presented in "Curso sobre Tecnicas y Estrategias para el Diseno de Opciones Tecnologicas como Parte de la Investigacion en Sistemas de Cultivo". Turrialba, Costa Rica: CATIE-IDIAP (Panama).

Netting, Robert McC. 1977. **Cultural Ecology.** Menlo Park, California: Benjamin/Cummings Publishing Co.

Orlove, Benjamin S. 1977. **Alpacas, Sheep and Men.** New York: Academic Press.

Raintree, J. B. 1984. **A Systems Approach to Agroforestry Diagnosis and Design.** Nairobi, Kenya: ICRAF.

Ruthenberg, Hans. 1980. **Farming Systems in the Tropics.** Oxford: Clarendon Press.

Schultz, Theodore W. 1964. **Transforming Traditional Agriculture.** New

Haven: Yale University Press.

Shaner, W.W., P.F. Philipp and W.R. Schmehl. 1982a. **Farming Systems Research and Development.** Boulder, Colorado: Westview Press.

Shaner, W.W., P.F. Philipp and W.R. Schmehl (eds). 1982b. **Readings in Farming Systems Research and Development.** Boulder, Colorado: Westview Press.

Spicer, Edward H. 1952. **Human Problems in Technological Change: A Casebook.** New York: Russell Sage.

Tax, Sol. 1953. **Penny Capitalism.** Washington DC: Smithsonian Institute.

Taylor, Paul S. 1976. "Walter Goldschmidt's Baptism by Fire: Central Valley Water Politics". In Loucky, J.P. and J.R. Jones (eds.) **Paths to the Symbolic Self: Essays in Honor of Walter Goldschmidt.** Los Angeles: Department of Anthropology, University of California, Los Angeles.

TAC. 1978. (Technical Advisory Committee, Review Team of the Consultative Group on International Agricultural Research.) **Farming Systems Research at the International Agricultural Research Centers.** Washington DC: World Bank.

Wallace, Ben J. 1984. **Working the Land: A Case Study in Applied Anthropology and Farming Systems Research.** Dhaka, Bangladesh: Bangladesh Agricultural Research Council.

Whyte, William F. and Damon Boynton. 1983. **Higher-Yielding Human Systems for Agriculture.** Ithaca: Cornell University Press.

Winkelmann, Donald and Edgardo Moscardi. 1982. "Aiming Agricultural Research at the Needs of Farmers." In Shaner **et.al.** 1982b. Boulder, Colorado: Westview Press.

Zandstra, Hubert C. 1982. "A Cropping Systems Research Methodology for Agricultural Development Projects". In Shaner **et.al.** 1982b. Boulder, Colorado: Westview Press.

PART II

CONCEPTS IN FARMING SYSTEMS RESEARCH

2

Anthropologist, Biological Scientist and Economist: The Three Musketeers or Three Stooges of Farming Systems Research?

Robert E. Rhoades, Douglas E. Horton, and Robert H. Booth[1]

Introduction

Every American movie-goer of the late 1950's remembers the contrasting images of the Saturday afternoon matinee. For comic relief were Larry, Curley and Moe, individualistically stumbling and banging their way through a job, getting in each others' ways, and always ending up with pie smeared all over their faces. Then came the serious three; efficient and coordinated musketeers with their blades of steel, swishing, pivoting and parrying to the final, clean victory in the face of enormous odds.

Teams. Although Farming Systems Research means different things to different people, one common ingredient is the interdisciplinary "team". Shaner **et.al.** (1982;185) write;

> Farming Systems Research and Development requires interdisciplinary teams - i.e., teams whose members represent different professions or disciplines. Although much is written about interdisciplinarity and its definitions, the key ingredient for true interdisciplinarity is **interaction.** This interaction invariably leads to synthesis and synergism.

It is largely assumed, as presented in on-farm "how-to-do-it" research manuals and funding proposals, that interdisciplinary teams resemble the "musketeers" more than the "stooges" and that the combination of anthropologist, economist, and biological scientist leads to "synthesis and synergism".

The authors of this paper belong to the three above mentioned disciplines and are long-standing members of successful interdisciplinary "teams" responsible for the generation and transfer of appropriate technology to thousands of Third World resource-poor farmers. However, it is our frank and sincere admission that our team most of the time resembles the Three Stooges at their best (or worst, however you choose to look at it). Instead of "synthesis" we spend an enormous amount of time in what we euphemistically call "constructive conflict" which at the moment it occurs can be deeply personal and difficult interaction. Although what we do only slightly resembles descriptions in FSR manuals, we keep doing it because we believe it is the basis of our success.

In this paper, we wish to informally tell our story[2]. In doing so, we hope to accomplish three goals; 1) explore the interdisciplinary "process" the way it actually occurs in an agricultural research institute; 2) describe how the separate disciplines perceive each other and why we feel that professional and institutional barriers make the chances of failure in achieving interdisciplinary research high; 3) examine means to overcome resilient disciplinary boundaries in order to get on with the business of generating appropriate and acceptable technology.

The Setting: International Potato Center (CIP)

The International Potato Center, with headquarters in Lima, Peru, is one of thirteen agricultural research centers which make up the Consultative Group of International Agricultural Research (CGIAR). The Center's research is organized around ten interdisciplinary "thrusts" ranging from collection and utilization of germ plasm to post-harvest technology and management. Most of the Center's interdisciplinarity involves biological scientists, given that only five of some sixty core staff positions are occupied by economists (4) and anthropologists (1). However, each social scientist works with biological science colleagues on at least one team. This paper deals largely, although not exclusively, with the interaction of anthropologists, economists and post-harvest technologists in the post-harvest thrust.

An understanding of the intellectual environment at a technical research center such as CIP is important for interpreting the tensions **and** creativity arising from intense, daily disciplinary contact. First, given that biological scientists dominate, there was an overall skepticism of the value of social science research, especially "pure" research which does not focus on the technology generation process. This is understandable given that biological scientists are under strong pressure (both self-motivated and institutionally fostered) to accomplish the technical goals of the center. Additionally, due to historical reasons, there is clearly a disciplinary prestige ranking which ranges from the plant breeder to agronomist, to economist, and now to the Johnny-come-lately anthropologist. The breeder claims a good variety will create its own suction force so why does he need an agronomist standing between him and the farmer. The agronomist argues that an "economically aware" agronomist can do farm budgeting so why does he need an economist. The economist, in turn, says he can make "social observations" so why anthropologists. Professional ethnocentrism is strong in agricultural research and it lends a strong "tribal" flavor to disciplinary interactions.

An Anthropologist Looks at Interdisciplinary Team Research

When I took up my job at CIP in 1979 my training in anthropology had not prepared me to cope with the special challenges of agricultural research. Graduate school had emphasized doing research alone, single authored papers, and an in-house orientation. The notion of doing research in "teams" with people from other disciplines was limited to summer archaeological digs. The graduate school role model at the time was Napoleon Chagnon who, at least in his early days, everyone idealized for his rugged individualism, his ability to do research alone among the "fierce people".

I have since concluded that, compared to team research at the potato center, Chagnon's situation was a piece of cake. When I arrived in my new "village" in 1979 tensions were running feverishly high among members of two interdisciplinary teams working on potato research in Peru's Mantaro Valley. The issue was not wife stealing but rather debates over approaches to on-farm research and technology design and transfer. Emotionally, however, I doubt if the adrenaline flowing was any less than among the Yanomamo at the height of a chest pounding duel. One team member warned me; "around here, its best to duck your head to about 3 feet off the ground, close your eyes, and keep running." Good advice.

Although I vowed early on to remain neutral, I found it impossible. There was a sort of ritualized form of "insults" which were passed back and forth between members of different disciplines. I was totally unprepared for the intense questioning that I as an anthropologist had to face. Although I had done some summertime AID-type consulting, I had in the large been around anthropologists who all agreed that anthropology was a Godsend. Suddenly I was out of the environment where anthropology was valued for itself. Instead, the issue was constantly forced on me; "prove what anthropology can contribute" or "why does agricultural research need anthropology?"

Answering these questions was not easy. At first I experienced great difficulty in expressing precisely what role anthropology could play. The old "farmer doesn't grow improved maize 'cause it makes lousy tortillas" argument fell on deaf ears. Little cultural tidbits seemed insignificant upside the "profits" and "yields" arguments of economists and agronomists. The second resentment or frustration I felt was that my role as an anthropologist was being defined by economists and agronomists, e.g. "what social factors are important in this agroeconomic farm trial?" I felt forced away from my holistic training, away from ecological and economic anthropology to a reductionist approach focusing on problems of interest more to the economist or agronomist than to me. The next three years were spent trying to define both for myself and for others what anthropology could contribute to agricultural research and development.

In the process, I found I had to deal with the needs and ambitions of economists, on the one hand, and biological scientists on the other. Despite common bonds and even the same department administrative affiliation, anthropologists and economists seemed to have had greater difficulty working together than either with biological scientists. Someone remarked cynically, but accurately; "Murder occurs more frequently within the family than outside of it."

In retrospect, the problem that anthropologists and economist have with each other probably arises less from subject matter competition than from problems of rewards and recognition. At least in my own case when I perceived that I might be destined for a service role to economics, I balked. It was better to form alliances with biological scientists than to become a "pet anthropologist" to an economist. Economists, on the other hand, had their own problem of becoming "pet economists" to agronomists or other biological scientists. Ruttan (1982) has remarked;

> Where social science staff has been cast in a purely service-oriented role, ...low staff morale and

difficulty in retaining an effective social science capacity had tended to result.

The intense questioning to which members of various disciplines subject each other is probably in the long-term functional since it forces an increasingly tighter focus on the job at hand. For example, when it became clear that ethnography as I had learned it in the university was inappropriate to applied agricultural research, I had to seek ways to modify traditional ethnography. Secondly, I had to explain anthropology so that it was understandable and relevant to biological scientists. Many excellent research anthropologists have failed in agricultural research simply due to their inability or unwillingness to modify anthropological methods and approaches (Rhoades 1983; Werge 1984).

The changes I (and Robert Werge before me) had to make were 1) drop the tendency to "study everything" and focus on the problems to be solved; 2) use ethnographic research to take concrete action in generation of technology or provide data useful for technology transfer or building national agricultural programs; 3) modify and blend methods into an interdisciplinary team approach and make findings relevant to non-anthropologists; 4) keep an eye on the broader relevance of a piece of research for purposes of extrapolation.

Another problem faced by agricultural anthropologists is that while they must constantly defend and prove the value of anthropology to biological science and economic team members, they must also face criticism from fellow anthropologists. Biological scientists are not interested in too much detail (pure research) while fellow anthropologists want more (and better "academic" level) research addressing theoretical issues in anthropology.

Many heated team arguments at CIP between anthropologist, economist and biological scientist were resolved over time as the team gained concrete experience with particular technologies. In other words, a discipline was vindicated or shown wrong through the technology generation process. In the end (never clear at the moment), we have come to realize that "everyone was right, and everyone was wrong."

Team research on post-harvest technology is a case in point. The direction of both potato storage and potato processing research has been vigorously debated at CIP. There have probably been more conflicts (personal, disciplinary, intellectual, etc.) in this "thrust" than any other in the center. This is due to the on-going team approach of the thrust. At times, team members through "constructive conflict" drive each other to extremes. However, it is this continuing conflict that continues to shed clear light on the technological problem at hand.

The next two pages trace through time the conflicts growing from the generation of post-harvest technology (potato storage and potato processing). Note that dialogue was continual and conflictive. Agreement comes hard and only after long time periods have passed. Social scientists were involved in different ways at all stages in technology generation, transfer, and impact. At many difficult moments there has been the temptation to pull back from the team and pursue disciplinary studies on the subject. This we feel often happens leading to "multidisciplinary" interactions but not interdisciplinary.

Chronology of design and generation of
post-harvest technology at CIP
Potato Storage: Case 1

Year

1975. Post harvest technologists work on "losses" in potato storage on experiment station.

1976. Anthropologist argues "losses" is scientist's notion. Farmers don't perceive "losses" as a problem and have no interest in ware (food) potato storage, rather seed storage of improved varieties.

1978. Team develops seed storage technology called Diffused Light Storage (DLS).

1979. Economist, who belongs to 'enemy' Production Constraints Team, argues, according to his partial budget analysis, it is not profitable to adopt DLS[3].

1980-81. Farmers adopt rustic stores and prove that economics is the 'dismal science'. Post harvest technologist and anthropologist feel vindicated. Economist forced to admit that anthropology may have some value.

1982. Funding agencies want to know what is the impact of DLS thus partial budget statistical data of economist suddenly become valuable. Economist happily vindicated. Post harvest technologist and anthropologist show more interest in economics.

1983. Post-harvest technologist questions earlier anthropological finding that ware potato storage not important. Conflict develops pitting technologists against both anthropologist and economist.

1984. Team research of anthropologist and technologist shows surprising amount of interest in ware storage. Anthropologist and post-harvest technologist reconcile difference while economist questions validity of their research findings.

Potato Processing: Case 2

Year

1976. Solution to potato processing problem in Central Andes seen by technologist as solar drying; builds "black box" to speed up drying.

1977. Anthropologist tests black box "on-farm" and shows it is unacceptable to farmers, arguing it is 10 times more expensive than traditional methods and no better.

1979. Technologist, guided by anthropologists "on-farm" information, re-orients project to developing labor saving processing equipment and a **papa seca** (dried potato) village level plant.

1980. Anthropologist/Sociologist shows in follow-up market study little demand for **papa seca** (consumed mainly in a once-a-year festive dish) and limited acceptance of plant.

1982. Technologist responds by developing more flexible plant and consumer oriented products; pursues "mix" aimed at urban poor.

1982. Anthropologist argues that "mix" may not be acceptable for economic, cultural, marketing, etc. reasons, based on study of slums and analysis of causes of previously failed "mixes". Technologists argue anthropologist went off on a tangent and conducted a longer, detailed study not relevant at this early stage of project development.

1984. Debate escalates and reveals clearly different perceptions of the problem from social scientist's and technologist's viewpoints; team now refers to potato "utilization", seeking creative ways (including mixes) to use potatoes (not only processing) which are not bound for fresh market.

1985. Research center adds sweet potato, a low cost crop amenable to processing, to research mandate, thus giving rise to heightened interest in post-harvest utilization. Technology developed on potatoes also becomes relevant to sweet potato processing.

These two cases illustrate that technology development did not occur in a straight line. Nevertheless, issues were confronted head-on and were normally resolved leading to a more profound understanding of the problem.

After almost ten years of struggle, anthropology along with economics has become accepted and institutionalized at CIP. Today, one almost never hears the challenge "prove what anthropology can contribute". The occasional challenge comes from biological science post-doctoral students fresh from university training with little experience in the complex realities of agricultural research. They, like most anthropology students, have not yet discovered the new agriculture.

An Economist's View on Working with Anthropologists and Biologists

Since 1973 12 anthropologists and 15 economists have worked at CIP as employees or visiting scientists (for brevity's sake, we now refer to discipline representatives as "ants", "cons", and "bios"). Most of these social scientists have worked on interdisciplinary teams with bios, but seldom have cons worked on teams with ants. Instead, each con and ant have tended to stake out an area of technological turf on which he or she could work with bios. In the few cases when ants, cons and bios have worked together on the same team, life has not always been easy. Professionals from these three disciplines have tended to identify and approach problems differently, use different methods to solve them, and use different criteria of success. Often like the 3 stooges, team members have seemed to spend more time and energy flailing around and arguing with each other than working efficiently together in problem-solving research.

Despite problems of interdisciplinary research, our experience leads us to believe that bringing ants, cons and bios together on teams enhances their contribution to the problem-solving process.

Problems of working with ants and bios

The first problem of interdisciplinary research encountered by CIP's cons was that many bios and ants did not want to work with them. Many ants and some bios felt that conventional economic models were irrelevant in the context of developing countries, and hence economics was of little value to an international agricultural research program. Many bios who took a less drastic stance still argued that economics had no useful role to play in generating technology, and that cons should keep themselves busy on narrowly defined "economic" problems, such as marketing. Among those bios who did appreciate the value of economic analysis of new technologies, some confused economics with accounting and assumed that anyone with common sense and a hand calculator could do the necessary calculations.

The lack of a generally accepted model for interdisciplinary team research led to conflicts over goals, approaches, resources, and methods. As a result, CIP's original large teams tended to splinter into smaller ones which competed with each other to prove that their own approach was better than the other teams' approaches. This competition and an inevitable conflict associated with it strained personal relations but stimulated a great deal of valuable innovation in interdisciplinary research approaches and methods.

Economics is Irrelevant In the mid-1970 s, when the cons and ants began to work at CIP, an argument voiced by many ants and some bios — particularly when results of an economic analysis indicated that new technology was less profitable than farmers' traditional practices — was that "western" economic theory and methods were not appropriate for developing country conditions. Hence, results of these economic analyses were irrelevant. Ants argued, for example, that since small Andean farmers were subsistence-oriented, micro-economic analysis based on the theory of the firm in a free market, was not useful for understanding peasants, their behavior or analyzing the economics of technological change. Cons, on the other hand, argued that ants did not appreciate the value of economics because **first,** they did not understand the concept of "opportunity cost", and **second,** they romanticized peasant life and underestimated the importance of small-farm production for the market.

This debate between cons and ants was not resolved until fieldwork involving both groups led to a common ground of experience and data. Ants and bios began to value the contribution of cons in the applied research setting as cons gained experience, accumulated examples of useful (and useless) economic analysis, and learned to appreciate the limits of their own discipline and the comparative advantages of other disciplines.

Economists should stick to "economic" problems A second, and very substantial, barrier to interdisciplinary research was the common misconception of CIP's bios that technological and socioeconomic problems are completely separate and can be solved independently. In practical terms this means that social scientists should concentrate on social and economic problems (e.g. marketing and extension) and leave the solution of technical problems to the biological scientists.

Based on this assumption, many bios argued that effective problem solving required a **sequence** of specialized tasks performed by experts, rather than **teamwork** involving different disciplines working together. An ideal sequence envisioned by the bios was something like this: **First** the con would survey an area to identify market potential for potatoes. **Then** a bio would identify and solve the key biological problems — for example design a new structure for storing surplus potatoes after harvest. **Finally** the con was supposed to solve the economic problem — determining where to build the store and how to sell the potatoes after storage. **At the tail end of the sequence**, and only in case of emergency, an anthropologist might be called in to study why farmers were not using the "improved" technology. Preference was given, of course, to the anthropological follow-ups that showed happy farmers using the technologists' pet technology. Only as experience and evidence were accumulated did bios begin to realize that "biological" and "socioeconomic" problems are inseparable, and must be solved together. This substantial change in "world view" led to much greater interest in interdisciplinary team research at CIP.

Horseback economics Many of CIP's bios considered economics to be little more than glorified accounting, and assumed that anyone with a little common sense and the ability to balance a bank statement could also do the economic analysis needed in an agricultural research institute. The handy, back-of-the-envelope calculations preferred by these bios are generally termed "horseback economics". In fact, they are usually more aptly described as "horseshit economics". The large number of unused, technically sound but economically unfeasible, potato stores dotting the landscape in developing countries stands as silent reminders to this fact[4].

Horseback economics has two essential problems. **First,** the types of calculations made are generally copied mechanically from an elementary farm economics or agricultural engineering text written for extensionists in the USA or Europe. These formulas are seldom directly applicable to semi-subsistence conditions in developing countries. **Second,** the prices plugged in by horseback economists are usually pulled out of the air, cooked up in Ministry offices, or jotted down in supermarkets or hardware stores. Consequently, they have little or no relation to the actual opportunity costs facing farmers, merchants, or governmental institutions which will ultimately decide to use or reject a new technology.

Lack of a model for team research In the absence of an accepted model to guide research teams made up of ants, cons, and bios, when CIP began farm-level research in Peru's Mantaro Valley, all members of the Social Science Department and several biological scientists began working together on the same loosely structured team which conducted preliminary surveys in the area. Soon, however, conflicts over goals, methods, and leadership resulted in a splintering of the original team into three smaller teams, with one social scientist and one or more biological scientists on each. Given CIP's decentralized organizational structure, these teams were free to develop their own interdisciplinary research models and methods.

Initially, many of CIP's bios questioned the value of having social scientists on these teams, and both cons and ants were under strong pressure to prove their worth. The cons and ants had very different ideas of what should be done: cons advocated quantitative analysis of production constraints based on questionnaire surveys, on-farm experiments, and cost-return studies. Ants, in contrast, argued that production problems could best be understood and solved through use of informal surveys and simple farm-level trials involving more interactions with farmers and less quantitative analysis. Both cons and ants wanted to "prove" they were right. Hence, rather than working together - which would have required compromise - they preferred to work alone.

Rebellion of the ants

CIP's first social scientist was a con, and the head of the Social Science Department has always been a con. Hence, ants are not only newcomers, but they are a kind of third class citizen, below both the bios and the cons. Especially in the early years, ants tended to feel that cons liked to keep them down, rather than share second class status with them. Consequently, in order to break out of their third class status and get "out from under" the cons, ants have been eager to establish strong ties with powerful members of CIP's bio-elite. This was an important reason why few ants expressed interest in participating on teams with economists. This situation was further complicated by a "divide and conquer" approach of some bios.

The challenge for economists

In the previous paragraphs we have discussed barriers to interdisciplinary research and problems a con is likely to face when working with ants and bios. Now let us turn to the legitimate challenges presented to economics by ants and bios and how CIP's cons responded to them.

Understanding the limits of economics Many cons assume, or at least act as if they do, that economics has the "final word" in assessments of farmers' technology needs and the process of technological change. In crudest form, they assume that if a new technology has a high rate of return farmers will adopt it, and if it doesn't, they won't. Of course, most published economic studies present sophisticated formulae and complicated reasoning. But this crude assumption lies at the heart of more economic analysis than most economists would like to admit. Interaction with ants and bios and frequent testing of hypotheses against actual experiences have helped CIP's cons appreciate the strengths and weaknesses of their theory and methods. We have learned the value of complementing our economic calculations with direct field observations and discussions with farmers. As a result, over time, we have given increasing emphasis to "informal" survey methods which get principal investigators into the field, and less emphasis to "formal" questionnaires which keep the principal investigators in the office, planning, supervising, and analyzing survey results. We have also moved away from complex experimental designs for on-farm trials, toward simple designs which can be readily understood by farmers, and hence, provide the springboard for fruitful interactions between investigators and farmers.

The value of "quick and clean" methods Many cons criticize the methods and results of ants as being "soft" — qualitative and subjective. But CIP's cons have learned that ants can generate a remarkable amount of useful information fast. In several instances long before the con completed his data processing, the ant finished his field report and moved on to another research task. This observation reinforced our move toward informal surveys, sharply focused single-visit surveys, simple experimental designs and other means of reducing the time required for fieldwork and data processing.

Somewhat to our surprise, speeding up the research process has not reduced the quality of our work. On the contrary, completing fieldwork and report-writing quickly has improved the overall productivity of applied research, because additional information and insights gained from frequent interaction with bios and farmers outweigh the loss due to simpler, less sophisticated data gathering and analytical procedures.

As a result of team interaction and constructive conflict, CIP cons now use simpler methods, pay closer attention to the relevance of their analysis (rather than its "scientific" rigor), and consider professionals of other disciplines as equal partners in the research process. Formerly, cons tended to act as if their training made them uniquely suited to be team leaders. We no longer believe this to be true.

Continuing problems

Despite progress made in interdisciplinary team research at CIP, many problems remain. One is that the structure of rewards and the pressure for quick results create strong tensions within and between teams, make them somewhat unstable, and encourage a strong bias toward research with a short-term payoff. A second problem is that CIP's relatively few social scientists are kept so busy with their work on technology teams that they have little time or no incentive to work together. Nor do they have time to "service" all the Center's important technology teams. Finally, within agricultural research institutes such as CIP, biologists maintain higher status and more authority than social scientists. As a consequence, there is a latent tendency to consider ants and cons as assistants to the technologists. In this role, which both ants and cons fiercely fight against, the actual contributions of social scientists to the applied research process is far below its potential.

The reward structure Among the International Agricultural Research Centers — all of which are applied research institutes — CIP is known for being sharply focused on priority problems and for achieving fast results. Consequently, CIP rewards scientists who are associated with the development of new research techniques or technologies which increase potato production in developing countries. In contrast, quality research which has no immediate application is not always encouraged. CIP's ants and cons are strongly motivated to work with bios on technology problems that appear to have quick solutions.

The pressures for success are so great that many bios support socioeconomic work as long as it casts their technology in a positive light, but oppose it when results are negative. For example, when CIP began research on means of using true potato seed (TPS) rather than seed tubers for planting material, there was strong support for economic research which showed that using TPS substantially reduced production costs/hectare. However, when on-farm trials began to show that using TPS also reduced yields, bios cried that it was "too early" to subject this new technology to economic analysis.

Being associated with an unsuccessful technology has several negative consequences for a social scientist at CIP. First, one foregoes the generally positive rewards which accrue to being on a winning team — which might include favorable publicity, travel and pay increases. Secondly, one's professional credentials and team loyalty might be called into question. And third, one might be denied the right to publish results which reflect negatively on the team's work. Such pressures are certainly not conducive to the conduct of the highest quality social science research.

Spreading the social sciences too thin In a research center like CIP, where there are at least ten biological scientists for every social scientist, once interest has been stimulated in interdisciplinary research, there simply aren't enough ants and cons to go around. Bios tend to view both ants and cons as multi-purpose "socio-economists". Hence, if a bio "believes" in the social sciences, he will want to try to get an ant or a con on "his" team, but seldom will he want, or could he get, two. As a result, there are extremely few opportunities for ants and cons to work together on the same interdisciplinary teams. Moreover, the small number of social scientists is in the institution and the numerous problems of serving as a member of more than one research team means that only a few of CIP's teams have social scientists as members.

A related problem is that, due to the small number of social scientists and the high demands for their services, they generally end up doing a number of jobs rather superficially. Consequently, much of their work is unpublishable. In the "outside world" publishing is important for professional recognition. Hence, there is a continuous tension between working "for the potato" and working "for your profession." The issue becomes particularly important when one enters the job market and is asked by prospective employers what he or she has done in the last five years besides helping design seed potato stores for small farmers.

A Biological Scientist Looks at Team Research with Social Scientists

FSR has to some extent become synonymous with interdisciplinary team research involving both biological and social scientists. In this section, I, as a biological scientist, wish to discuss the needs for, or justification for such interdisciplinary team research, and to discuss the problems of developing and maintaining such a team. The reader should be aware from the outset that I am a biological scientist who accepts the value of socioeconomic involvement in technology transfer. Hence, the views expressed may be less critical than those of some of my disciplinary colleagues.

Traditionally, agricultural research has been disciplinarily organized and executed. This commonly led to the situation where biological scientists kept themselves busy in agricultural research stations doing research **about** production problems. When attempts, through extension agencies, were made to transfer the experiment station results to farmers they were commonly faced with the response "We could have told you so" by their social science colleagues no matter whether the transfer attempt was successful or otherwise. Being more specific, the anthropologists could be heard to say "Obviously farmers in the region/country X are/are-not going to be interested in that type of technology" while the economists would pronounce that "Our partial budget analysis clearly demonstrated that the technology would/would-not be adopted". Both genre of social scientists seem to be low risk takers. They are not willing to stick their necks out with a technology but always seem to hedge waiting to see which way the wind shifts. They frequently prefer to restrict their research to low-risk descriptive activities.

Two major challenges face farming systems or interdisciplinary team researchers; 1) getting anthropologists and economists involved in the A to Z of technology generation and transfer in agricultural research and development projects and not simply in initial surveys or after-the-fact evaluations; and 2) getting the biological scientists to accept that improving farmers' practices requires more than technical solutions.

The importance of constructive conflict

In facing these challenges and building truly interdisciplinary research teams, I believe that a continuing phase of "constructive conflict" between the potential team members is **essential.** The continuing constructive conflict phase is necessary to develop disciplinary understanding and respect between individual team members. Team members must develop a confidence in the potential contribution of the other disciplines and individuals.

As the name implies, the constructive conflict phase is not an easy one. It not only involves intra-team dialogue, frequently conflictive during its early stages, but can also create an impression of "continually arguing" or "trouble making" amongst colleagues not actively involved in the team. Similarly, it cannot be considered a once-and-for-all process but rather needs to continue through the whole process of problem solving. If the constructive conflict phase is by-passed, so as to avoid the "embarrassment" of "conflict", little team spirit develops and only what may be termed **multi**disciplinary **groups**, and not truly **inter**disciplinary **teams** are formed.

For constructive conflict to take place successfully a

suitable institutional framework must exist which permits and promotes interdisciplinary, not disciplinary, arrangements and uses of resources (human, physical and financial). It is, for example, unlikely that successful constructive conflict will occur in a large "team" as is commonly found in FSR. In such cases, the large "team" should be broken down into smaller functional units in which constructive conflict can occur and assist to more closely define the problem solving objective of the unit.

Seen from the biological scientist's point of view, tensions in team research often arise in the early stages of interdisciplinary dialogue due to: 1) the anthropologist's "know-all" attitudes to the reality of the farmers' situation but common lack of knowledge concerning the technologies which make up the farmers' situation; and 2) the economist's frequently superior planning attitude and his attempts to certify his superiority through the use of little understood economic jargon. Economists have greater difficulties than the biological scientists in both communicating the principles of their discipline to non-economists and to applying and using these principles at the farm level. Economists often complain about their low status, a fact which perhaps reflects "sour grapes" due to economics' loss of control of agricultural research management in the wake of the "Green Revolution". However, despite a general control of top level management by biological scientists, economists continue to have great influence over policy and receive excellent rewards for their effort. Their "low status" is a figment of their imaginations.

The main advantage of the constructive conflict phase is the bringing together of disciplines, and their respective principles relative to the problem to be solved, into an interdisciplinary team capable of developing a potentially acceptable biological/technological solution to the farmers'/consumers' defined problem. The process clearly involves learning or in-service training element for all the team members during which the potential role of each discipline is understood and accepted. The jargon surrounding each discipline is either removed or understood, and some elementary knowledge of some of the basic principles of all the disciplines is acquired. This results in development of respect for the potential contribution of each discipline and individual. Clearly, in the formation of small closely-knit teams it would be foolish to forget personality effects.

Team leadership and balance of viewpoints

Another important process which occurs during constructive

conflict is arriving at a correct balance between the disciplines involved and their relevance to the problem to be solved. This is a difficult and sometimes painful process. One of the major aims of all agricultural research institutions is the generation and transfer of improved agricultural technologies. Thus all disciplines involved in any teamwork should focus on the given technology. This is not, however, the same as saying that the biological scientist should also be the dominant member of the team. For various technical and personal reasons it may be more appropriate for other disciplines to play a dominant role. For example, it may be that credit limitations or social customs limit the transfer of a developed technology, in which case the economist and anthropologist, respectively, should play the dominant role.

Regardless of who is required to play the dominant role, attention should not be allowed to drift away from the team's objective, and the potential technological solution. All team members must service the defined objective and not be tempted into side studies unless these are agreed to be necessary to the accomplishment of the objective. Anthropologists appear to be particularly prone to the above, and, if not controlled, are commonly tempted into complex and complete studies of particular communities or situations. While such a study may be interesting, it may go well beyond what is required for the generation and transfer of the specific agricultural technology which is the objective of team efforts. Similarly, with the economists' apparently natural leaning towards planning, there is commonly a tendency for economists to attempt to dominate the team and impose or over-state the importance of that discipline in accomplishing the team's objectives.

The arrival, through constructive conflict, at the correct disciplinary balance is crucial to the success of the team's effort and as the balance may need to change with time so the process of constructive conflict will need to continue.

Once a team is formed it is not easy to keep the team together. Difficulties often focus around the necessary partial loss of individual and disciplinary identity. Perhaps the easiest way to elucidate these difficulties is by making an analogy with the development and running of a sports team. A sports team may be made up of a group of highly qualified individuals (disciplines) each trying to maximize personal achievement, or it may consist of a group of team players intent on working together to achieve a common goal. In this context, the first sports team is comparable with a multidisciplinary group and the second to a truly interdisciplinary team. Also, if the constructive conflict phase has not been successfully completed problems of team leadership and authorship of publications can arise. Within a small team which successfully manages constructive conflict such problems rarely arise and, the actual contribution of each discipline will normally dictate publication authorship and "leadership" which might well change over time.

The main advantage of keeping the team together clearly lies in the ability to maintain overall momentum and focus of interest with a higher degree of probability of achieving the defined technological objective. Once the objective is reached the team is disbanded, and members are reassigned to new teams according to the specific new research problem. Such a "new" team attacking a "new" problem needs to repeat the process discussed above, even if it is comprised of the same team members. The inevitable termination of team activities can lead to misunderstanding on behalf of non-team members; "What happened to the interdisciplinary team then, fallen-out have we?" Such misunderstanding of "constructive conflict" and the "disbanding" of teams adds to the difficult but rewarding task of interdisciplinary team research. Again perhaps we should make an analogy with sport; either you prefer individual sports or you prefer team sports. I think the same is true in agricultural research and it is well that scientists and administrators recognize these, among other, differences if the limited human resources are to be used to best advantage.

Conclusion

In this paper, members of an interdisciplinary team - an anthropologist, economist and biological scientist - have explored their experiences in team research. The three sections were written independently and have been left largely in their original form. An important message of the paper - hopefully relevant to new team members - is that "conflict" is an important ingredient in successful research. It is better for practicioners, however uncomfortable it may seem, to pursue the conflict than to take the comfortable way out of drifting into disciplinary isolation.

Although tongue-in-cheek, we have drawn an analogy with another team, the three stooges. It is our hunch that most successful teams experience rough going. This to a great degree grows from the difficult nature of on-farm research activities. Combining disciplines, executing on-farm experiments, functioning under the sometimes ludicrous conditions of underdeveloped regions, reconciling personality differences, writing joint papers, etc, are all difficult tasks which can be disastrous. This, as we have noted, is what drives teams often to the "multidisciplinary" approach, wherein team members pursue disciplinary research and pass back and forth jargon-ridden, mutually incomprehensible reports. It is passed as interdisciplinary research and it satisfies donors, but it is not the real thing. Combining disciplines to arrive at a broader understanding of farm conditions and then generating technology to solve problems is not so easy.

It is our feeling that in the interdisciplinary process, especially on the post-harvest team, "new" fields of agricultural anthropology, economics and biological science have been created. While we maintain disciplinary integrity, our understanding of agriculture has broadened and become more holistic compared to the narrow, compartmentalized research of each discipline. This interdisciplinarity, in turn, will help in the development of an alternative agriculture for the modern world.

Notes

1. Anthropologist, economist and technologist at the International Potato Center, Lima, Peru.

2. Each of us wrote our sections independently without consulting the other. We have purposely left the original text more or less as it was, even at the risk of repetition.

3. The economist co-author of this paper responds to this statement by noting that the partial budget analysis in question was actually done by an agronomist. The dangers of this type of "horseback economics" are discussed in the next section.

4. The technologist co-author of this paper responds to this statement by noting that in fact elaborate economic and marketing analyses were frequently made by professional economists before the "white elephants" were constructed. Many of the stores were not only technically sound but were shown on paper by economists to be capable of making a profit. Nevertheless, they remain empty. The problem really comes from direct transfer of packaged technologies without an adequate understanding of either the food production/demand pattern into which they are being placed or the existing post-harvest practices.

Bibliography

Rhoades, Robert. 1984.
Breaking New Ground: Anthropology in

Agricultural Research. Lima, Peru: CIP.

Ruttan, Vernon. 1982.
Agricultural Research Policy. Minneapolis, Minnesota: University of Minnesota Press.

Shaner, W.W., P.F. Philipp, and W.R. Schmehl. 1982.
Farming Systems Research and Development. Boulder, Colorado: Westview Press.

Werge, Robert. 1983.
"The Uses of Ethnography in Agricultural Development in the Andes: Anthropologists of the International Potato Center". Paper presented at a symposium in honor of William Carter at the Annual Meeting of the American Anthropological Association. Nov. 19, 1983. Chicago, Ill.

3

Basic vs. Applied Research in Farming Systems: An Anthropologist's Appraisal

Stephen B. Brush

The theme of this paper[1] is the relation between basic research in anthropology and applied agricultural research known as Farming Systems Research (FSR). The role of anthropology in Farming Systems Research has been a topic of much recent interest (DeWalt 1985; Tripp 1985). This paper approaches the relation differently by examining the importance of FSR for anthropology and asking whether certain basic anthropological research can be better done in the context of an applied institution. It concludes that while basic anthropological research can be improved through more cooperation with applied research, impediments created by institutional frameworks and pressures must be acknowledged in order to strengthen both sides. Differences between methods and short-term objectives constitute acknowledged constraints, but these differences also suggest more general institutional tensions between basic and applied research.

This paper explores the cooperation of two basic anthropology research programs with an applied social science program at the International Potato Center (CIP) in Peru. The case is valuable because of the commitment of CIP to anthropological research and because of a similar commitment by anthropologists to do basic research connected to CIP's interests. In spite of very good conditions for collaborative research between anthropologists and other social and agricultural scientists, the collaboration was only partially successful. This paper hopes to contribute to improved collaboration between basic and applied social science by examining the success and failure of this case.

Basic and Applied Research in Anthropology

The distinction and relation between basic and applied science is common to all science and has been an important issue in anthropology for the last forty years. Basic science refers to research whose principal objective is to test and to advance theoretical propositions and generalizations. This work is largely carried out by university based scholars, and the products of the work are publications in scholarly journals. Applied science refers to the use of scientific knowledge in contexts where testing general theory is not the objective. It may rely on the theoretical underpinnings of basic science, but its method and rationale are not validated by the academic canons of basic science, theory testing, generalization and publication. It is concerned with problem solving and with specific situations rather than with testing general propositions. An active dialogue between the two sides has been an important objective for many years under the assumption that the health of both sides would benefit (Foster 1969; van Willigen 1984).

In anthropology, **basic science** elucidates the nature of culture by studying such phenomena as cultural adaptation and evolution, the structure of social systems, and the logic of symbolic systems. Divisions among schools, (e.g. materialist, structural, linguistic and symbolic anthropology), topical specialties (e.g. economic anthropology, cognitive anthropology) and specific cultural area groups (e.g. Andean, South Asian) reflect the normal science of the discipline. The overall objective to learn how culture influences human behavior is fulfilled by bringing ethnographic case study material to bear on general questions. The gathering and interpretation of ethnographic material utilizes several methods, the most important being participant observation and cross cultural comparison.

Priorities for basic research may be set within the general schools of anthropology but they are more often controlled by more specific research communities and culture area groups. Above all, it is important to recognize the academic context of basic research in anthropology. The overwhelming majority of basic research careers in the discipline are lived out at universities and colleges. Priorities in research therefore reflect the academic measures of success, especially publication and research grants. These, in turn, depend on how well a specific project relates to general issues, that is, how well it commands and articulates with published literature in the field and how well it promises to generalize from specific case study research. The measure of basic research projects must often be long rather than short term, because of the time needed to process case study material and relate it to a much larger field of literature. A result is that evaluation criteria and procedures are rather amorphous and often ambiguous even for specific research projects. Academic reviewers are usually anonymous and specific research projects **per se** are seldom reviewed in academic performance evaluations. Basic researchers in anthropology usually view their work as autonomous and insist that the proof of success can only be judged by the quality of publication and teaching.

In anthropology, **applied science** is usually concerned with the design and implementation of public programs and social policy; e.g. education, health, housing, and rural and urban development. Van Willigen (1984) identifies information, policy and action as products of applied research. The theoretical base of anthropology contributes to the role that anthropologists play in public institutions concerned with these issues. That base emphasizes four aspects of anthropology: 1) the role of cultural factors in determining social phenomena (the holistic aspect), 2) the importance of understanding the local perspectives (the "emic" view), 3) the value of case studies, and 4) the utility of participant observation for analyzing human behavior and social systems.

Apart from the general theoretical and methodological underpinnings that applied anthropology takes from the more basic science, its objectives, priorities, and routines are very different. Objectives and priorities are set within and by a specific institution rather than the academic and amorphous context of basic anthropology. Short term results are expected, in contrast to university programs. The evaluation criteria and procedures are usually explicit, and publication is not desired as a primary product of the research. Anthropological methods that characterize basic research are not usually understood or appreciated. Participant observation is very time consuming and may not generate data with which others in the institution are accustomed to work. Cross cultural comparison may be seen as a diversion from the specific task at hand. Unlike basic research, applied research projects are generally evaluated according to their contribution to the institutional goals. This means that anthropologists doing applied work in institutions do not act as autonomous scientists, but as members of larger teams. They are not responsible to a general and ambiguous system but rather to a concrete and explicit institution and set of actors who participate in the same research program. In this context, the anthropologist's work is not judged by other anthropologists but by scientists and managers who may hold very different assumptions about social theory and methods.

As in many other sciences, basic and applied agricultural researchers share an uneasy but important relationship. They work in different contexts, with different objectives and methods, and with different measures of success and failure. Basic researchers are concerned with the failure of applied scientists to rise above specific situations and to publish their work in a way that pushes theory forward. Applied scientists, on the other hand, may criticize the irrelevance of much basic research, arguing that it is inaccessible to non-specialists and inappropriate to the design and implementation of public policy.

The Basic Research Component in FSR

During the last decade, a new approach to international agricultural research known as Farming Systems Research (FSR) has been adopted by many research and development centers (Gilbert, Norman and Winch 1980). This research method is defined by several characteristics: a) the integration of social and biological scientists in single teams, b) an emphasis on farm level surveys for setting research agendas, c) on-farm testing throughout the research program, and d) reliance on adoption of technology as a measure of success. FSR is promoted as a way to reach small farms where improved technology has not been adopted, to generate appropriate technology, and to increase farmer participation in the development process (Shaner **et. al.** 1982). It is also perceived as a type of applied agricultural research that is compatible with basic science (Simmonds 1984; Sands 1985). Anthropologists have been proponents of FSR and participated in it because of its emphasis on farm level data and anthropological research methods.

The antecedents of applied FSR lie partly in the basic social sciences, especially economics. More importantly, the origins of FSR lie in the structure of applied agricultural research programs and the desire of social scientists to find a more legitimate role in institutions dominated by biological or production scientists. A key institution in this history is the international agricultural research center (IARC) established in the 1960s and 1970s to harness agricultural science for development (Ruttan 1982). From their inception, IARCs were dominated by the production sciences, with social science relegated to a minor role. The success and impact of the technology generated from these centers was a compelling object for social science research. The IARCs turned to social science for several reasons. Social science was needed to refute the claims that IARC technology was biased toward wealthy farmers and actually worsened the plight of peasants and landless laborers. The high adoption rates of new rice and wheat in Asia were not shared by other commodities or by all poor farm regions. It was hoped that social scientists could ameliorate this by identifying constraints to adoption and by helping to develop outreach and extension programs. The FSR program emerged simultaneously with social science programs at IARC's, and social scientists there were quick to recognize FSR's utility. An example of this is the early development of FSR at CIMMYT (the International Maize and Wheat Improvement Center) (Byerlee **et. al.** 1982; Tripp 1985).

In spite of the origins of FSR in agricultural economics, other social scientists, especially anthropologists, saw roles for themselves in FSR (DeWalt 1985). Anthropologists' assets for FSR included their knowledge of the small farm sector in underdeveloped regions, their methodological skills in doing research among non-Western cultures, and their holistic approach to identifying farmer criteria for the selection of new technology. Besides these practical skills, anthropological theory provided a relevant backdrop, especially the interest in the economics of subsistence systems. As DeWalt (1985; 111) notes, natural continuity exists between well established interests in cultural ecology, economic anthropology and FSR.

Andean Anthropology

Research on Andean agriculture has progressed significantly during the last decade. A major organizing theme of Andean anthropology has been the idea of "verticality". Stimulated by the work of John Murra, a broad front of research was undertaken in the 1970's around this concept (Salomon 1982). Murra (1975) defines verticality as an Andean ideal to achieve self-sufficiency by the exploitation of multiple agricultural zones along an environmental gradient. The environmental gradient provides a physical and mental template for the organization of this world. This is most clearly true for the agriculture of the region, but is also true for local and regional economies and political systems and for rituals and cosmologies (Platt 1976). This concept suggests that many Andean entities, from households to states pattern their behavior according to the vertical landscape of the Andean world.

The research stimulated by the concept of verticality led to more complete descriptions of the actual land-use patterns

found in the Andes, for example an improved description of the production zones that define Andean agriculture (Guillet 1981). These provided a stronger sensitivity to the legacy of Andean culture in determining contemporary life in the area. An example of this is the discovery of the persistence of such practices as community controlled crop and field rotation systems (Mayer and Fonseca 1979). The concept also opened a lively debate on the nature of Andean peasantry that weighed local features in relation to more generalized features of precapitalist economic systems (Lehman 1982; Painter 1984). Finally, verticality research put Andeanists in a better position to ask detailed questions about the nature of agricultural ecology in the area. Two topics for continued research within this general framework were 1) how production zones were determined in relation to environmental, social and economic factors, and 2) the success of Andean agriculture in managing a mountain agroecosystem. These topics were the point of departure for research between Andeanist anthropologists and CIP in 1977-1978.

Production zones

After the initial interest sparked by the discovery of the replication of vertical land use patterns in many different Andean contexts, anthropologists focused on the dynamics of production zones. They asked how these zones were defined, how they varied, how they affected agriculture, and how they changed. These questions went far beyond the scope of the original verticality theme. Ethnographic data and more recent economic history replaced the ethnohistoric focus of the original verticality research.

An ambitious research program on the organization of Andean agriculture around production zones was initiated by Enrique Mayer and Cesar Fonseca in 1974 in the Canete Valley, south of Lima (Mayer and Fonseca 1979). Their basic proposition was that specific elements of Andean social organization create conditions for stability and security of production in the face of environmental diversity and risk. This is accomplished through the mechanism of production zones comprised of crops, animals, land tenure, water rights, and the organization of labor and the agricultural calendar. Communal forms of organization played an especially important role (Mayer 1985).

In order to explore this agroecological organization and to test propositions of the role of communal versus individual economic decisions, Mayer and Fonseca began a transect of the Canete Valley from the coast to its watershed divide with the interior Andean valley to the East, the Mantaro Valley. A first step was to map the production zones of the valley. Ten zones were defined and mapped, from intensive agro-industrial zones at sea level to communally controlled potato zones above 3,000 meters. Key questions about these zones concerned the exercise and conflict of individual and communal control over the resources in each zone, and the evolution of zonation under different economic and political conditions. Important influences here were demographic changes and the rise of commercial production in different zones (Mayer and Fonseca 1979). Traditional Andean land use patterns, such as community controlled sectoral fallow appeared to degrade as commercial replaces subsistence agriculture.

Managing the Andean agroecosystem

A second topic in anthropological studies of Andean agriculture to emerge from the verticality idea was the success of traditional technology in coping with and managing the precarious mountain environment. The success of Andean agriculture was apparent in the prehistoric civilization of the area, in the large population depending on relatively simple technology and the apparent lack of serious land degradation. Geographic and anthropological research suggested a number of specific adaptations to reduce risk (Gade 1975; Brush 1977), with agriculture playing an important role. The assertions of risk management and environmental stability in Andean agriculture were, however, largely based on conjecture rather than systematic investigation, with no comprehensive model of the risks in the Andean agroecosystem other than casual observation on stability or degradation in the Andean environment.

Agricultural zonation and the distribution of crops were frequently cited in models of cultural adaptation in the Andes (Gade 1975; Mayer 1985). Crop diversity between and within production zones corresponded to the vertical diversity of the region. Only limited research by anthropologists or other cultural ecologists on specific crops had been done on the dynamics of crop selection. The potato (**Solanum** spp.) was a logical candidate because of its importance and history in Andean agriculture. It was originally domesticated in the Andes, and great genetic diversity of the crop had been found there. Research on potato agriculture provided a way to more precisely describe stress, response and stability in the Andean agroecosystem.

My initial research on Andean agriculture had two

objectives. The first was to describe its cultural foundations. Questions under this objective dealt with how farmers identified and selected potatoes, and how genetic diversity in the crop was related to knowledge and other cultural factors. The primary methods to achieve this objective were a study of the folk taxonomy of Andean potatoes and a study of the relation between folk and botanical classifications of potatoes. This research wished to establish the consistency in naming and selection as an important factor in the distribution of potato germplasm. The second objective was to describe the exchange and distribution of potato seed. Questions under this objective dealt with the maintenance of seed stocks, practices to maintain these, and transactions between farmers and regions that affected potato germplasm distribution. The methods to achieve this objective were socioeconomic surveys on seed systems within and between villages.

Convergence with CIP

In 1977, the research programs on production zones and on managing Andean agroecosystems converged with the research program of the International Potato Center (CIP). The projects on production zones and agroecosystem management were originally conceived as basic research programs, but both were selected partly because of their relevance to economic development and agricultural change. Although anthropologists are regular members of international development projects and agencies, their success at CIP has been unusually high. Prior to these two anthropological projects, CIP had benefited from Werge's successful research on post-harvest technology (Werge 1983) and from Eastman's (1977) rural sociology of potato production in Peru. Rhoades (1984) attributes CIP's openness to anthropologists to both the institution and the anthropologists who have worked there. Institutional qualities include flexibility, pragmatism and openness toward social science beyond economics. Anthropologists inside the institution have contributed significantly to the design of appropriate technology, especially in post harvest technology and to understanding the food systems of which potatoes are a part. Qualities of the visiting social scientists include a willingness to work within a single commodity framework.

Mayer and I were ready to continue our research in 1977, as CIP was undertaking a major new socioeconomic study in central Peru on potato production, to identify constraints and to evaluate alternatives. The Mantaro Valley Project was to be CIP's entry in the newly formed Farming Systems Research effort, to develop techniques and expertise in on-farm research (Horton 1984). CIP, however, never used the phrase Farming Systems Research nor the formal terminology of FSR, such as "recommendation domain" to describe this work. Three technical areas for interdisciplinary research were identified; agronomic constraints to production, post-harvest technology (storage and processing), and seed production and distribution. The project was supported by an external grant from Canada's International Development Research Centre and drew together a staff of social science and agronomy research assistants under the direction of Douglas Horton. Four general methods were employed: a) a reconnaissance of the study region to define agroecological zones, b) a single-visit farm survey to identify farm types and production constraints, c) a multiple-visit farm survey to describe farm management, and d) on-farm trials of different technological packages. Separate but related research on post-harvest technology and on potato seed production was carried out concurrently (Werge 1977; Monares 1981).

For the baseline study on the agricultural ecology of the Mantaro Valley, CIP turned to Enrique Mayer, who was completing his transect of the Canete Valley. Mayer was a native of the Mantaro Valley and knew it intimately. His Canete transect ended at the valley's borders, and his trans-Andean transect would eventually cross it. CIP's idea for a survey of the Mantaro valley fit nicely with his own research agenda. Mayer and a research assistant mapped three zones (high, middle and low altitude) and several subzones within these (Mayer 1979). The primary characteristics of these zones and subzones were their physical features relating to altitude, precipitation and the cropping pattern within each, especially the percentage of land under different crops. Other important characteristics included whether the crop was grown for subsistence or commercial purposes and the degree of intensification in agriculture. Mayer's survey of the Mantaro Valley for CIP used roughly the same units of analysis and question relating to production zones as used in the Canete research.

The Mantaro Valley reconnaissance corroborated some of the general conclusions from the Canete Valley study. The household and peasant community related to one another in similar fashion, and the dynamics between subsistence production and commercial production were analogous in both areas. For Andean studies, an important contribution of the Mantaro study was a more complete description of a regional farming system than any found in earlier regional studies, including the Canete study. Zonation related not only to environmental factors but also to location relative to roads, markets and urban centers.

Differences from the Canete study were the greater significance of market production by peasant households living in communities and the large numbers of small holders and landless laborers not associated with peasant communities. These make the simple dichotomies between subsistence and market economies and between household and community control much more difficult to sustain.

For CIP, Mayer's land-use map identified zones and types of producers and was a key product for organizing subsequent research in the area and for targeting research elsewhere. A critical discovery in Mayer's study was the importance of an intermediate production zone (3,500-4,000m) which produces as many potatoes as a lower zone (3,000-3,500m) but under peasant farm conditions. This zone became the focus of the Farming Systems project to identify production constraints on different sized farms.

I had independently developed the theme of the Andean potato agriculture project, but I had redesigned it after consultation with the Center to align my work with that of the Social Science Unit. In particular, the project was shifted from a study of the impact of improved varieties to an examination of how farmers identify, select and distribute potato varieties and seed tubers. Principal funding for the project came from the National Science Foundation and the College of William and Mary. CIP agreed to provide additional funding plus logistical support. CIP's primary interest in this research was to gain knowledge of the native seed production and distribution system, especially among small farmers. Potato seed is a costly and critically important component in potato production, and little was known about how small farms handled it.

An important issue was the relation of the cultural system to the biology of the crop. This required collaboration between myself and geneticists and taxonomists at CIP, a novel type of collaboration at CIP. This collaboration was intriguing to the geneticists and taxonomists who had basic science interests in the biology and evolution of the crop, although my work was unconnected to their applied breeding work.

Geneticists familiar with Andean potato farmers knew about native lexicons for potatoes but assumed that these were neither systematic nor consistent. They also expressed the conviction that the genetics of the potato were best understood as a purely genetic issue and that the inclusion of farmer knowledge and behavior would only muddle the picture. These assumptions and convictions had to be overcome during the research, to build the necessary collaboration with geneticists.

Disagreement between CIP's Social Science Unit and my research was evident at the outset. To CIP, the folk taxonomic work was overly academic and not connected to their specific interests in seed systems. To me, it was the essential starting point of a larger research program. Questions of seed flows and selection were to follow the initial work on describing how Andean farmers identified and classified their numerous potato varieties. A **modus vivendi** between my project and the Mantaro Valley Project was established when I agreed to do my research in the same geographic area. Tension continued, however, over my desire to work in the higher altitude zones. The goal of describing Andean agricultural adaptation stressed the importance of diversity, and the greatest diversity in potatoes is found in relatively high elevations (above 3,500m), and it is associated with small farms and subsistence production. The high production zone thus became my primary focus, but it was of only marginal interest to the CIP project. CIP focused on the intermediate zone for several reasons. Greater agricultural change and commercialization were found there in comparison to the higher zone, and this zone was more accessible to the research staff. CIP also believed that on a population basis the intermediate zone was more representative of the total sample of the region's potato farmers. Their research efforts, consequently, neglected the highest zone where traditional Andean agriculture was still practiced.

My research provided a description of a folk taxonomic system with great complexity at the varietal level, with new data on the complexity of Andean agriculture at the level of the household and field (Brush **et.al.** 1981). It indicated that the folk naming system had internal consistency and could be related to genetic classification of the potato, at least within limited geographic areas. It also provided a basis for understanding the selection of varieties by peasant farmers, pointing out their primary criteria for evaluating new varieties. This was especially important for understanding the adoption of improved varieties. This research also showed how peasant farmers maintain genetic diversity of potatoes. Finally, it described the traditional system for distribution of potato seed across altitudinal and regional boundaries, although this received less attention in resulting publications.

For CIP, this research had three results. First, it contributed an "emic" dimension to the framework for understanding the selection and adoption of potato varieties. This emic dimension included the peasant rationale for maintaining diverse collections. Second, it facilitated communication and cooperation between social scientists and geneticists, especially between myself and two geneticists at CIP, Zosimo Huaman and Carlos Ochoa. It helped define a research question of interest to both: how the behavior of peasant farmers affects the germplasm of domesticated potatoes. This question is relevant not only to understanding the dynamics of the crop but also to planning genetic conservation. A third positive consequence was in helping to define a new area of research for the Social Science Unit, the adoption of new potato varieties.

Several important differences existed between the two basic anthropology projects. Unlike Mayer's project, the field methods of mine had not been refined by earlier research. Another difference was that expected results of Mayer's project were specified, while mine were not as clearly defined. Finally, Mayer's entire budget was supplied by CIP, whereas I received important logistical and scientific support but only minor financial support from the Center. A negative consequence of these differences was that my research did not fulfill CIP's expectations for a study of the potato seed industry in the Mantaro Valley. The emphasis on folk taxonomies and selection criteria required the majority of my time, and my work in the higher altitude zone limited my interaction with the Mantaro Valley Project, working in the lower zones. From my perspective, my research succeeded by compiling a unique data set for future publication. From CIP's perspective, it failed to produce a descriptive study of seed flows in the Mantaro Valley. This was in comparison to Mayer's project that had produced a report published by CIP. The success and failure of these two projects within CIP's applied Farming Systems Project offer interesting insights into the limits of integrating basic anthropological research into FSR programs.

Compatibility between Basic and Applied Agricultural Research

The very attempt of two anthropologists to carry out basic research programs in conjunction with applied research suggests a certain level of compatibility. In judging this, it is important to recognize two factors. First, CIP's contribution to the overall research program of the anthropologists was only partial. In Mayer's case, his Mantaro Valley reconnaissance for CIP was only one segment of a more ambitious program to extend a socioeconomic and agroecological transect across the Andes. For me, the research on folk taxonomies and seed networks were only initial steps in a larger program on Andean agroecology. Second, our contribution to the overall Mantaro Valley Project was limited. Mayer's involvement ended with the reconnaissance. Although resident in the study area and in frequent contact with the Mantaro team, I did not participate directly in the design or conduct of the Mantaro Valley Project. This non-participation was troublesome to both sides, and the difference in research agendas was never fully resolved.

One notable contributing factor to compatibility between CIP and the two outside anthropologists was the interest of CIP's Social Science Unit in recruiting outsiders. Important here was the willingness of CIP's Social Science Unit and other scientists at CIP to support basic anthropological work that went beyond the specific goals of the Farming Systems project. Both of the outside anthropologists understood their contributions to CIP's work as an important but limited part of their own research. CIP's willingness to support the pursuit of goals that it did not set contributed to the collaboration but also was a source of friction. CIP recognized that outsiders could contribute to their programs, and the outsiders recognized the benefit of the expertise in both social and agricultural sciences at CIP.

Another factor was the anthropologists' acceptance of the critical importance of potato production for understanding Andean economic systems and cultural ecology. Such crop-specific research is rare for the anthropology of other agricultural systems. Both anthropologists benefited from the tremendous concentration of expertise on potato agriculture that is found at CIP, as well as the challenging give-and-take with social scientists engaged in applied research.

A third contributing factor to this compatibility was a complementarity between the anthropological research and the Center's programs. The anthropologists offered a knowledge of the region and of Andean agriculture not available at the Center. CIP offered a technical base in potato agriculture and strengthened the anthropological research. For instance, the Center's germplasm collection and expertise in the biological dynamics of the crop were essential components of the folk taxonomy research. In many instances, the single commodity focus of agricultural research and the holistic expectations of anthropologists make collaboration difficult. In this case, the interdisciplinary scope of CIP actually complemented the holistic perspective of the anthropologists. The important agreement was that the potato needs to be viewed as both a cultural and a biological artifact.

Two Problems

Beyond the basic collaborative framework that tied the

visiting anthropologists to CIP's applied Social Science Unit, the mutual development of common research thrusts did not occur. Mayer's primary concern was the importance of the community in controlling Andean agriculture. Mine was the cultural base of Andean agroecology, exemplified by the identification and selection of diverse potato varieties. Neither of these themes, however, contributed significantly to the Mantaro Valley Project.

The divergence in emphasis between CIP and the anthropologists doing basic research indicates a fundamental demarcation between farming systems and basic anthropological research. Two assumptions of the farming systems project drew it apart from the anthropological research. The first concerned the relevance of social organization above the household. The second concerned the relation between change and tradition. This demarcation arose from different views of objectives and methods in applied and basic research that were discussed at the beginning of this paper.

The role of the community

Both the farming systems economists at CIP and the anthropologists agreed that several components of the farm system were relevant, but they differed on the significance of these, and this difference was clearly reflected in their respective research designs. Both implicitly agreed that the production system of the Mantaro Valley was composed of individuals, households, farms, communities, urban and industrial centers, public and private services, and commodity, labor and factor markets.

For the anthropologist, the role of the community signifies the importance of Andean culture in determining the economic behavior of the region. Anthropologists have noted the ubiquitous presence of communal organization since pre-Hispanic times and their importance in organizing Andean agriculture and economy (Murra 1975). Communities have been significant in defending the Indian population against domination and exploitation (Mallon 1984). Contemporary communities control important aspects of the rural economy of the region. They own large portions of the land, determine the agricultural calendar, and manage sectoral fallow systems, communal pastures, irrigation systems, and the transportation infrastructure of the rural economy (Guillet 1981; Mitchell 1976). Communities are particularly important in establishing the framework within which the household operates (Brush and Guillet 1985).

What is not clear is the relation between the household and the community in the overall shape and direction of the farm economy. Many Andean scholars believe that community involvement is a **sine qua non** of development. The community controls strategic resources (e.g. land and water) and provides essential services (e.g. road maintenance). Whereas it appears that subsistence production is heavily influenced by the community, for instance the agricultural calendar, commercial production seems less under the community's influence (Guillet 1981; Mayer and Fonseca 1979). Commercialization is clearly a major factor in agricultural change in the region. It is, however, erroneous to assume that the only change in the rural economy derives from increased commercialization. Change also results from such other factors as increased population pressure on standard field rotations and from the unavailability of labor because of off-farm employment.

Systematic research on the farm economy of the region should, therefore, include the community as a significant component. This was missing in the design of the Mantaro Valley Project. Although that project recognized farm system components above the household, it undertook no research on their influence in household decision making. Mayer's reconnaissance indicated that community control was most pronounced in the higher zones. Although CIP was interested in the problem of community control, its selection of research sites in the lower zones discouraged further investigation along this theme. This was further discouraged by a research format that emphasized household decision making on the adoption of technology. The end result was that community-level variables played an insignificant role in the Mantaro Valley Project.

CIP, however, could not expect to do everything that might have interested the various team members or associates. The work on adoption of new technology was innovative and demanding. Moreover, the primary objective of the Mantaro Valley Project was not to advance Andean studies but to describe small farm behavior **vis-a-vis** technology adoption. Lessons learned were not relevant solely for Peru's farmers but for CIP socioeconomic research in other regions and countries. Their neglect of community variables, therefore, was understandable in this wider context.

Tradition and change in agriculture

The second assumption that divided the FSR Mantaro Valley Project from anthropological research was their differing perspectives on tradition and change in Andean agriculture. An implicit but serious rift between applied and basic research surrounds this issue, dividing scientists who define their immediate mission as how to change farming systems from those whose mission is to build fundamental knowledge abut the nature of those systems. The issue of tradition as an obstacle or object of study straddles this rift. From an applied Farming Systems point of view, the study of tradition is inappropriate in several ways. It diverts attention from and may not contribute to the immediate task of inducing change. It also requires methodologies that are questionable from an aggregate and econometric perspective. Applied FSR deals with populations and agroclimatic regions rather than cultures. Behavior and patterns that interest anthropologists because they typify a cultural tradition may not, in the economists' eyes, be representative of a region's population. Anthropologists, on the other hand, are not accustomed to selecting research topics or sites according to their representativeness. A benefit of collaboration in interdisciplinary teams, such as an FSR team, is to encourage different social sciences (such as anthropology and economics) and agricultural sciences (such as breeding and pathology) to be more rigorous in selecting sites and people to interview.

On the surface, anthropologists may occupy the role of defending tradition in opposition to FSR as the promoter of change. It is, however, erroneous to view the anthropological interest in tradition as opposition to change. Change and tradition cannot, of course be logically separated. Tradition does not denote anachronistic or illogical behavior by people who ought to know better. It is, rather, a reflection of the persistent selection of behavior and traits from the past in preference to "modern" ways. Tradition is a result of rational decision making and selection, and it is, therefore, as "modern" as any behavior. The study of change is an inherent goal of anthropology, manifested in such major theoretical areas as cultural ecology and political anthropology. Unfortunately, change is often only indirectly or implicitly studied by anthropologists. This is worsened by their reliance on functionalist or neo-functionalist models. Anthropology has struggled against ahistorical and homeostatic tendencies for four decades, and this fact indicates the tenacity of a disciplinary bias to emphasize structure rather than process.

From an anthropologist's perspective, it is proper to study a farm system by establishing the logic of traditional behavior and traits. Most FSR projects, however, do not accept this as a legitimate activity. FSR is designed to minimize the opportunity to analyze traditional patterns in order to move as quickly as possible to testing new technology. The well known reliance on the "sondeo" or rapid rural assessment in FSR exemplifies this design flaw (Beebe 1985). The rapidity of this method, its focus on a single commodity and its view of the farm household as an autonomous economic unit greatly limit its utility for the thorough, holistic approach desired by anthropologists. On the other hand, the limited and time consuming case study methods of the anthropologist are costly and usually do not adequately address the issue of sample representativeness except qualitatively. Methodological problems posed by the rapid rural assessment technique have been discussed by Beebe (1985), DeWalt (1985) and Tripp (1985). Although a good case is made for this technique, basic anthropology and other social science research depend more on exhaustive and lengthy methods. Middle-term research strategies of two to four years are accepted for production sciences in agricultural research institutions, and social sciences should insist on support for comparable strategies to supplement the initial rapid assessment techniques. The success of the Mantaro Valley Project was primarily due to the fact that they followed a rapid rural assessment with a longer term study of farm practices in the area, including intensive and repeated surveys of a carefully selected sample. These strategies might fruitfully combine the research programs of permanent staff and outside scientists. To build more effective middle-term strategies, social science units in agricultural research institutions need to mature beyond their initial service role mentioned above.

Conclusion

The Mantaro Valley Project provided a good context to view the problems of integrating basic anthropological research into applied FSR. While both applied and basic research were successfully conducted, several limiting factors constrained the full integration of the basic anthropological projects with their applied FSR counterparts. Cooperation was facilitated by the success of applied anthropology at CIP and by the interest of outside anthropologists in studying a single commodity. Integration was impeded by differing perspectives on the importance of the community and on the value of studying tradition. Communication was not helped by the pressure on all three projects to finish within relatively short field seasons and by the logistics of planning where CIP's central office was far removed from the field station. One lesson was that communication between different types of social science may often be more difficult than between social and biological sciences. The differences separating anthropologists and economists at CIP were seen in many areas: the definition of the research problem, methodology, and views of different types of evidence.

In spite of the constraints in integrating the Mantaro Valley Project with concurrent anthropological research, the general level of cooperation between CIP and these independent research anthropologists remained high. My project opened a new area of interest in the Social Science Unit on the importance of potato varieties in farmer selection. Both Mayer's and my work has been used as a foundation to subsequent CIP research on farmer behavior and Andean agriculture. Our work for CIP, in turn, has become a foundation for further anthropological research on Andean agriculture that emphasizes change.

The contributions of anthropology to applied FSR has been documented elsewhere (e.g. DeWalt 1985). For the health of basic anthropology in agriculture, the applied context of FSR can be equally beneficial by providing interdisciplinary focus and access to important data to test theory. This benefit, however, seems to be rarely taken by anthropology, partly because of the unfamiliarity of scientists doing basic research with the activities and resources of applied research institutions. As this case shows, there are other obstacles to overcome. Both production and social science research done in applied agricultural research programs are characterized by differences in objectives, methodologies, and evaluation procedures that separate it from basic science. These differences are exaggerated for social science by the service role that it plays in most international centers. As DeWalt observes (1985), it is imperative that greater attention be paid to the sociology of agricultural research.

The potential contributions of FSR to anthropology lie in both practical and theoretical areas. Perhaps its most important contribution is the theoretical and methodological links that FSR opens for anthropologists studying the cultural ecology of agriculture. Another contribution is that it obligates anthropologists to redefine holism. Although anthropologists pride themselves on the holistic nature of the discipline, sociocultural factors are only parts of agricultural systems, and anthropological holism alone is insufficient to cope with agricultural systems. In joining FSR teams, anthropologists must sharpen their craft to collaborate with economists, agronomists and other agricultural scientists. For instance, the issue of representativeness of samples assumes greater weight on such teams, and holism must shift from sociocultural to agricultural holism. Seen from a disciplinary perspective, this shift may appear to narrow our focus from general categories (e.g. culture) to more specific ones (crops). Agroecosystems are, however, comprised of both human and non-human factors, and any systematic research on them must integrate these two sides. Ecological anthropology has recognized this, but links to multidisciplinary research are difficult to establish and maintain. FSR provides an appropriate context to establish these links in agroecosystem research, although anthropologists must be aware of FSR's limits.

The contribution of FSR to anthropology should go beyond providing jobs for applied anthropologists. Anthropology can benefit from the interdisciplinary climate of FSR programs. Ecological and economic research in anthropology can benefit from the greater precision that an exhaustive study of a single commodity demands. Single commodity research is a new direction for anthropologists, but it builds easily on our interest in the nature of traditional farming cultures. Anthropologists are accustomed to perceiving culture embodied in key rituals or institutions. A farming culture can just as easily be seen embodied in a particular crop. We have known for many years that crops are a product and a symbol of cultural evolution. By asserting this fact we offer important insights to agricultural research, both basic and applied.

Recommendations

1. The principle of collegiality is essential to combining basic and applied research, especially when visitor and core scientists are engaged together. Each must respect the other's objectives and constraints, and this involves establishing and nurturing long-term interest in the research program of the other. Collegiality depends on sharing common objectives, but it also needs the support of institutional arrangements, such as a memorandum of understanding or cooperative agreement that will back-up both the visiting and host scientists representing basic and applied research.

2. The difference between grants and contracts governing research agendas needs to be clearly articulated and understood in the collaboration. Basic research is accustomed to the former and applied research to the latter. A combination of both types of funding modes builds the strongest relation between visiting and host, basic and applied scientists. Specific expectations must be stated and negotiated early in the collaboration, preferabley prior to research.

Notes

1. This article is reprinted here with the permission of the editors of **Human Organization,** where it first appeared in its completed form in Volume 45 #3. The research was supported by the National Science Foundation (BNS76-83067), the College of William and Mary, and the International Potato Center. I wish to express my thanks to the staff of the International Potato Center for their assistance and continued support of my research. I first organized my thoughts for this paper as a contributor to a symposium on "Farming Systems Research: The Integration of Social and Biological Approaches to Agricultural Development" organized by Jeffrey R. Jones and Ben J. Wallace at the 1984 Annual Meetings of the American Anthropological Association. I am grateful to Douglas Horton, Benjamin Orlove, Bill DeWalt, Peter Ewell, Gregory Scott, Lovell Jarvis, Charles Brown and Margaret Brush for comments and suggestions. The views expressed here are my own.

Bibliography

Beebe, J. 1985. **Rapid Rural Appraisal: The Critical First Step in a Farming Systems Approach to Research.** Networking Paper No. 5. Gainesville, Florida: Farming Systems Support Project, Institute of Food and Agricultural Sciences, University of Florida, Gainesville.

Brush, S.B. 1977. **Mountain, Field and Family: The Economy and Human Ecology of an Andean Valley**. Pittsburgh: University of Pennsylvania Press.

Brush, S.B. and D. Guillet. 1985. "Small-scale Agro-Pastoral Production in the Central Andes." **Mountain Research and**

Development 5(1): 19-30.

Brush, S.B., H.J. Carney and Z. Huaman. 1981. "Dynamics of Andean Potato Agriculture". **Economic Botany** 35(1): 70-85.

Byerlee, D., L. Harrington and D. Winkelmann. 1982. "Farming Systems Research: Issues in Research Strategy and Technology Design." **American Journal of Agricultural Economics** 64: 897-904.

DeWalt, B.R. 1985. "Farming Systems Research". **Human Organization** 44(2): 106-114.

Eastman, C. 1977. **Technological Changes and Food Production: General Perspectives and the Specific Case of Potatoes.** Lima: International Potato Center.

Foster, G.M. 1969. **Applied Anthropology.** Boston: Little, Brown and Co.

Gade, D. 1975. **Plants, Man and Land in the Vilcanota Valley of Peru.** The Hague: Dr. W. Junk B.V. Publishers.

Gilbert, E.H., D.W. Norman and F.E. Winch. 1980. **Farming Systems Research: A Critical Appraisal.** Rural Development Paper #6. East Lansing, Michigan: Michigan State University.

Guillet, D. 1981. "Agrarian ecology and peasant production in the Central Andes". **Mountain Research and Development** 1(1): 19-28.

Horton, D.E. 1984. **Social Scientists in Agricultural Research: Lessons from the Mantaro Valley Project, Peru.** Ottawa, Ontario: International Development Research Centre.

Lehmann, D. (ed). 1982. **Ecology and Exchange in the Andes.** Cambridge: Cambridge University Press.

Mallon, F.A. 1984. **The Defense of Community in Peru's Central Highlands: Peasant Struggle and Capitalist Transition, 1860-1940.** Princeton: Princeton University Press.

Mayer, E. 1979. **Land Use in the Andes: Ecology and Agriculture in the Mantaro Valley of Peru, with Special Reference to Potatoes.** Lima, Peru: Centro International de la Papa, Social Science Unit.

——. 1985. "Production Zones." In **Andean Ecology and Civilization.** S. Masuda, I. Shimada and C. Morris (eds). Tokyo: University of Tokyo Press:

45-84.

Mayer, E. and Fonseca. 1979. **Sistemas Agrarios en la Cuenca del Rio Canete (Departamento de Lima).** Lima, Peru: ONERN.

Mitchell, W.P. 1976. "Irrigation and Community in the Central Peruvian Highlands". **American Anthropologist** 78(1): 25-44.

Murra, J.V. 1975. "El Control Vertical de un Maximo de Pisos Ecologicos en la Economia de las Sociedades Andinas". In **Formaciones Economics y Politicas del Mundo Andino** by J. Murra. Lima, Peru: Instituto de Estudios Peruanos.

Painter, M. 1984. "Changing Relations of Production and Rural Underdevelopment." **Journal of Anthropological Research** 40: 271-292.

Platt, T. 1976. **Espejos y Maiz: Temas de la Estructura Simbolica Andina.** Cuadernos de Investigation CIPCA, No. 10. La Paz, Bolivia: Centro de Investigacion y Promocion del Campesinado.

Rhoades, R.E. 1984. **Breaking New Ground: Agricultural Anthropology.** Lima, Peru: International Potato Center.

Ruttan, V.W. 1982. **Agricultural Research Policy.** Minneapolis, Minnesota: University of Minnesota Press.

Salomon, F. 1982. "Andean Ethnology in the 1970's: A Retrospective". **Latin American Research Review** XVII(2): 75-128.

Sands, D.M. 1985. **A Review of Farming Systems Research.** Paper Prepared for Dr. Alexander Von Der Osten, Executive Secretary, Technical Advisory Committee/CGIAR.

Shaner, W.W., P.F. Philipps, and W.R. Schmehl. 1982. **Farming Systems Research and Development: A Guideline for Developing Countries.** Boulder, Colorado: Westview Press.

Simmonds, N.W. 1984. **The State of Farming Systems Research.** Washington, DC: Agriculture and Rural Development Department of the World Bank.

Tripp, R. 1985. "Anthropology and On-Farm Research." **Human Organization** 44: 114-124.

Van Willigen, J. 1984. "Truth and Effectiveness: An Essay on the Relationship between Information, Policy and Action in Applied Anthropology". **Human Organization** 43: 277-282.

Werge, R. 1977. **Potato Storage Systems in the Mantaro Valley Region of Peru.** Lima, Peru: International Potato Center.

———. 1983. "The Uses of Ethnography in Agricultural Development in the Andes: Anthropologists at the International Potato Center." Paper presented at the Annual Meetings of the American Anthropological Association. Chicago, Illinois.

4

Developing Notional Technologies in a Farming Systems Research Context

James Chapman[1]

Introduction

Recently there has been a growing emphasis on the development and diffusion of agricultural technology in less developed countries as a vehicle for rural development as well as for the provision of basic food supplies to the rural population. A great deal of investment in agricultural research has taken place, with major expenditures made by philanthropic foundations and developed country governments to fund the International Agricultural Research Centers (IARCs).

Partly as a result of this investment, high yielding and early maturing varieties of rice and wheat have been developed and diffused. While production technologies based on these varieties have had a large impact on total output, the derived benefits have not been distributed equitably among all producers. The technologies were developed without regard to specific sizes or types of farms. However, by choosing the types and levels of inputs of the agroclimatic conditions under which the technologies were developed, researchers in effect developed technologies appropriate for adoption by producers with command over high-quality natural resources as well as sufficient capital to obtain the necessary complementary inputs.

As numerous studies have demonstrated, the technologies of the Green Revolution were much less attractive to producers lacking the required resources (e.g. Morss **et.al.** 1976; Pearse 1980). The adopting group tended to be relatively large, well educated and wealthy farmers, while the non-adopters tended to be small-scale, uneducated, resource poor farmers. Social scientists have spent much time and effort in **ex post** evaluation of the reasons why small farmers have not benefited greatly from new agricultural technology. The major point that this chapter seeks to demonstrate is that it is possible for social scientists to join forces with biological scientists in the **ex ante** design and evaluation of technology consistent with the needs and circumstances of specific target populations (e.g. small farmers). Specifically, this chapter focuses on the development of **notional technology,** first defining the concept and then presenting an illustration of how the concept was used in a Farming Systems Research project in the Philippines.

Notional Technology Defined

As far as this author can determine, the term notional technology was coined by Anderson and Hardaker (1979), defined as scientifically undeveloped technology having a recognized potential for adoption by farmers in specific ecological and socioeconomic circumstances. More specifically, it presents the potential solution to one or more problems which inhibit the realization of gains in production or productivity, and gains in monetary and nonmonetary benefits to small farm families. It requires the creation of new concepts, or the modification of existing concepts. As such, the development of notional technology is as much an art as a science. It depends largely on the ability of researchers to analyze, synthesize and invent. According to Anderson and Hardaker (1979);

> Notional new technologies are, because of their hypothetical natures, cheap to invent and bounded only by the imagination of the inventor. Since more fully developed technologies usually have their genesis as notions, attention to generating notional new technologies should not be disregarded. Evaluation of this category can range from intuition, to analysis, but analytical appraisal is essentially confined to work with models rather than on real systems.

Talking about notional technology in an abstract sense does not provide the reader with a very clear picture of the process by which it is generated or its eventual usefulness. Therefore, the remainder of this chapter is dedicated to the presentation of an example where a notional technology was "invented" and evaluated during the course of a research project undertaken in the Philippines[2].

Project Background and Description

Research on cropping systems at the International Rice Research Institute (IRRI) was begun in the late 1960s by Richard Bradfield. Bradfield studied techniques for fitting a variety of legumes and other crops between rice plantings, with the primary objectives of improved human nutrition and soil fertility maintenance (1972). Through his experiments he revealed opportunities available for more intensive and diversified cropping.

In the early 1970s, research emphasis shifted from determining productivity of new or improved cropping patterns to the study of cropping patterns on existing farms where rice was a basic crop. In 1974, the Cropping Systems Program (CSP) was enlarged to include a multidisciplinary team to undertake research on existing and improved cropping patterns.

The CSP chose to focus efforts on **rainfed** lowland and upland rice areas in South and Southeast Asia. Priority was given to areas where it was possible to increase cropping intensity, i.e. the number of crops planted per growing season on a single unit of land.

The CSP concentrated on resource utilization on small rice farms, seeking to increase the benefits derived by crop production from available physical resources (e.g. rainfall, solar radiation, and soil) that are not readily modifiable (Zandstra 1978). It also considered biological and economic factors at the farm level as they influence the performance of cropping systems. Though CSP research was carried out in specific sites, the objective was the development of technologies, including new ways of combining crops into cropping patterns, which were appropriate for a large number of areas with similar climatic and physical conditions. Therefore, factors in the community or on the farm which restrict the adoption of new techniques did not necessarily force the abandonment of research on those techniques. A large part of the CSP's on-going effort was focused both on the generation of component technology[3] for cropping patterns and on management of improved technology. The generation of new component technology depended upon feedback from CSP researchers to biological scientists at IRRI[4].

The Project Setting

In 1975, the CSP began research in a rainfed rice growing area in Iloilo Province, Panay Island, West Visayas, Philippines (Figure 4.1). The research site was originally selected for its agroclimatic representativeness in terms of soil, water management, weather and geomorphic land relationships.

From 1975 to 1979, the CSP carried out the following activities at the site;

1. In early 1975, a baseline survey was taken of about 25% of all farmers in the site area (241 respondents). From baseline information, 45 farmers were chosen at random to participate in an intensive farm record-keeping study. Much of the information for the site description was provided by the selected farmers, who also provided a source of feedback for cropping systems researchers.
2. A large number of new cropping patterns were developed and tested on farmers' fields. The majority of the effort was focused on developing means of increasing cropping intensity (the number of crops planted in a single season), with special emphasis on growing two or more crops of rice.
3. Farm-household record-keeping activities were undertaken which involved mainly two aspects; (a) the collection and analysis of input-output data from experimental cropping pattern test fields; and (b) the collection of input-output data and prices to determine relative pattern profitablilities and resource flows in the farm-household economies of the 45 farmer cooperators.

The objective of the cropping system technology development was the utilization of high-yielding rice varieties with low photoperiod sensitivity. Varieties possessing this characteristic mature in a more or less constant period of time, regardless of day length. Traditional rice varieties generally mature during the same period each year, each variety responding to its own particular day length requirement. A traditional crop is often in the field for six to eight months before harvest. Modern varieties have tended to be of the early to intermediate maturing type. The principal advantage of early maturing varieties (EMVs) lies not in an increase in yield potential or pest resistance, but rather in the suitability for multiple cropping systems where more than one crop is grown sequentially on the same land during one year (Harwood

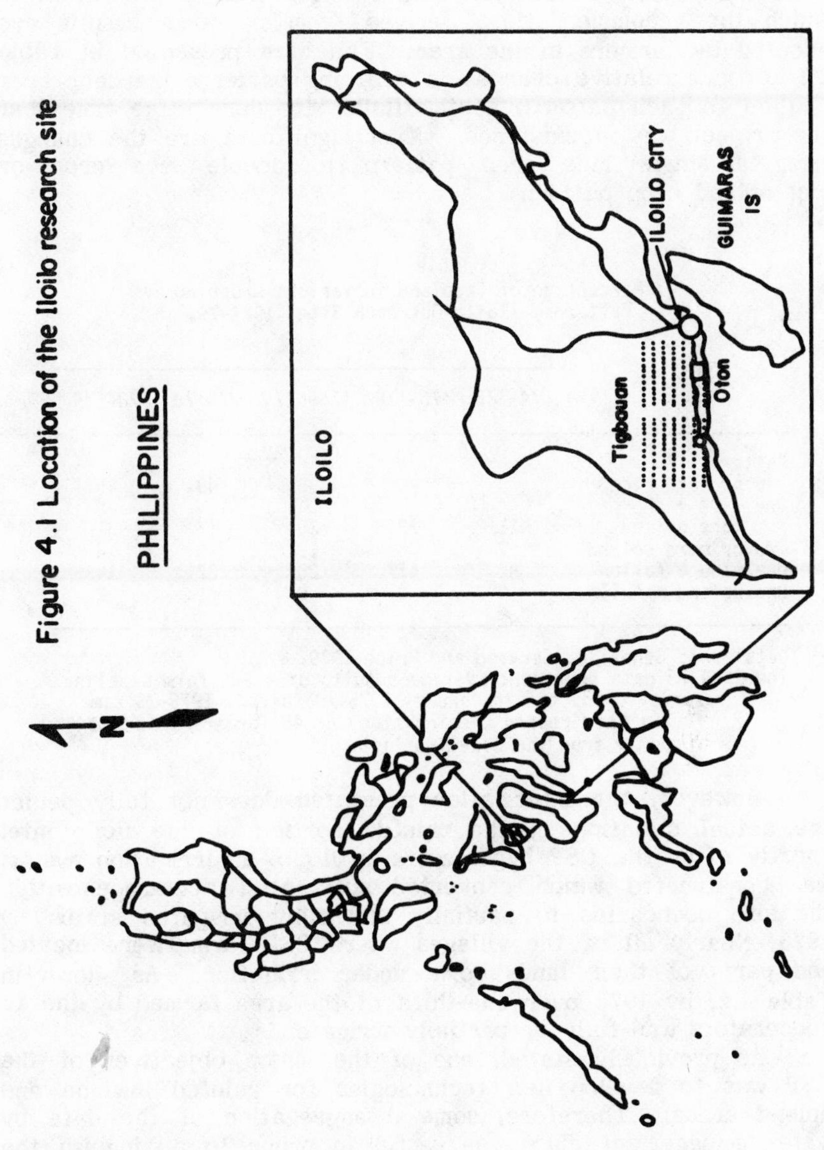

Figure 4.1 Location of the Iloilo research site

(Harwood 1976).

A good measure of the degree of success of cropping systems research carried out in a specific area is the extent to which the recommendations derived from research results are adopted by farmers in the area. The data presented in Table 4.1 present relative changes in cropping patterns (percent area planted to each pattern) at the Iloilo site during the time that the project was in existence. Most significant are the changes from a single rice crop pattern to double rice crop or rice-upland crop patterns.

Table 4.1
Percentage of Cropland in Various Cropping Patterns, Iloilo Outreach Site, 1974-79.

	'74-'75	'75-'76	'76-'77	'77-'78	'78-'79
Pattern					
Two or more rice	5	20	38	49	45
One rice + one or more upland	11	28	30	17	31
Two or more upland	2	5	12	6	8
One rice + fallow	82	47	20	27	14
One upland + fallow	0	0	0	1	2

Data from: Genesila, Servano and Price 1979.
The 1974-75 data represent average results of a 205 farm baseline survey conducted in January 1975. Data from 1975-79 came from a farm record keeping study on 45 farmers selected randomly from the baseline list.

However, the information presented does not fully depict the actual situation in the rainfed portion of the Iloilo site. Shortly after the CSP began work in Iloilo, an irrigation system was constructed which converted substantial hectarage within the site boundaries to partially and fully irrigated status by 1976. Nearly all of the villages where CSP farms were located had parts of their lands come under irrigation. As shown in Table 4.2, by 1978 over one-third of the area farmed by the 45 cooperators was fully or partially irrigated.

As previously stated, one of the major objectives of the CSP was to develop new technologies for rainfed lowland and upland areas. Therefore, some disaggregation of the data by water management class was useful in order to distinguish the effects of the research on the target population. In Table 4.3, the percentages of land cultivated by the 45 CSP economic cooperators devoted to different cropping patterns are displayed according to water management category. The figures demonstrate that technology focused on facilitating the establishment of two or more rice crops during a single growing season has been rapidly adopted by farmers with irrigated and partially irrigated land. The adoption rate of multiple rice cropping on rainfed lowland was much lower (19 vs 30%). Evidently, the double rice crop pattern for rainfed land was less stable because of year to year variations in rainfall intensity and duration. On the other hand, multiple cropping with one rice crop followed or preceded by one or more upland crops greatly increased in rainfed areas. Much of the increase was due to the fact that farmers adopted the early maturing varieties (EMVs) in the rainfed areas, thus increasing the amount of time that sufficient moisture would be available for planting other crops before and after rice.

Table 4.2
Percent of the total area of the farms of 45 economic cooperators under different water management classes, Iloilo, crop years 1976-79.

Crop Year	Irrigated	Partially Irrigated	Rainfed Lowland	Rainfed Upland	Total
1976-77	15	7	68	10	100
1977-78	24	5	61	10	100
1978-79	23	11	56	10	100

Source: Genesila, Servano and Price 1979.

The importance of water and the utilization of EMVs in facilitating increased cropping intensity is clear. All of the study farmers were in relatively similar positions in 1974 with respect to water management, since none of the area was irrigated. Those farmers with land that came under full irrigation managed to muster sufficient labor, power and material resources to enable them to plant two or more rice crops beginning in 1976. This suggested that water was a key resource limiting the adoption of more intensive cropping practices, especially with regard to planting two or more rice crops in a single season.

Table 4.3
Percentage of Cropland of 45 Farmers in Various Cropping Patterns by Water Management Category, Iloilo Outreach Site, 1976-79.

Cropping Pattern	1976-77[1]	1977-78[2]	1978-79[2]
		Rainfed Lowland	
Two or more rice	30	19	8
One rice + one or more upland[3]	43	30	46
Two or more upland	4	2	0
One rice + fallow	21	48	25
One upland + fallow	2	0	1
		Rainfed Upland	
Two or more rice	0	0	0
One rice + one or more upland[3]	66	8	6
Two or more upland	21	73	74
One rice + fallow	0	6	0
One upland + fallow	13	13	20
		Partially Irrigated	
Two or more rice	39	90	71
One rice + one or more upland[3]	0	3	22
Two or more upland	0	0	0
One rice + fallow	61	7	6
One upland + fallow	0	0	1
		Irrigated	
Two or more rice	96	100	96
One rice + one or more upland[3]	0	0	3
Two or more upland	0	0	0
One rice + fallow	4	0	0
One upland + fallow	0	0	1

1. Derived from Roxas and Genesila 1977.
2. Source: Genesila, Servano and Price 1979.
3. Upland refers to all crops grown in the area other than rice.

Cropping Potential in Iloilo

A somewhat clearer picture of cropping potential in Iloilo is presented in Figure 4.2. Time, in terms of months, is measured along the axes, while rice and upland crop growing seasons are depicted on the upper and lower portions respectively of the diagram. The rice-growing (drought-free) season usually lasts for five to six months (July through November). During the remaining months, the probability of drought stress conditions for rice is quite high. Upland crops can be grown during two periods of the year; at the beginning of the wet season (May to mid-June), and in the transition period from an overly wet to an overly dry climate (mid-October to mid-February). From mid-February through April, upland crops in the field would likely suffer from drought stress, while from mid-June to mid-October, the probability of excessive moisture and flooding is high. The "competitive" period, when either rice or an upland crop can be grown, generally has a duration of just over one month (mid-October to late November).

The season given for planting and growing rice and upland crops can be considered "safe" in the sense that in most years yield reductions will not occur during those periods due to either drought stress or excessive moisture. Planting and/or harvesting outside those two periods involves increased risks with the level of risk increasing as crops spend more time outside of the safe periods. One of the major objectives of cropping systems research is to design cropping patterns that will fit climatic conditions as closely as possible in order to protect farmers from unacceptable risks.

Farming Systems Dynamics in Iloilo: A Qualitative View

The preceding description of Iloilo farming conditions focuses very heavily on the description of crop agroclimate. The nature of farming systems is determined by social, economic and cultural factors as well. Such factors, however, do not lend themselves well to easy description. Rather than list important socioeconomic aspects of Iloilo farm systems here, an attempt is made to give the reader an idea of the interplay between society and climate by looking at a specific case.

The following narrative is based upon the situation and experience of an actual farm family. However, since there is a

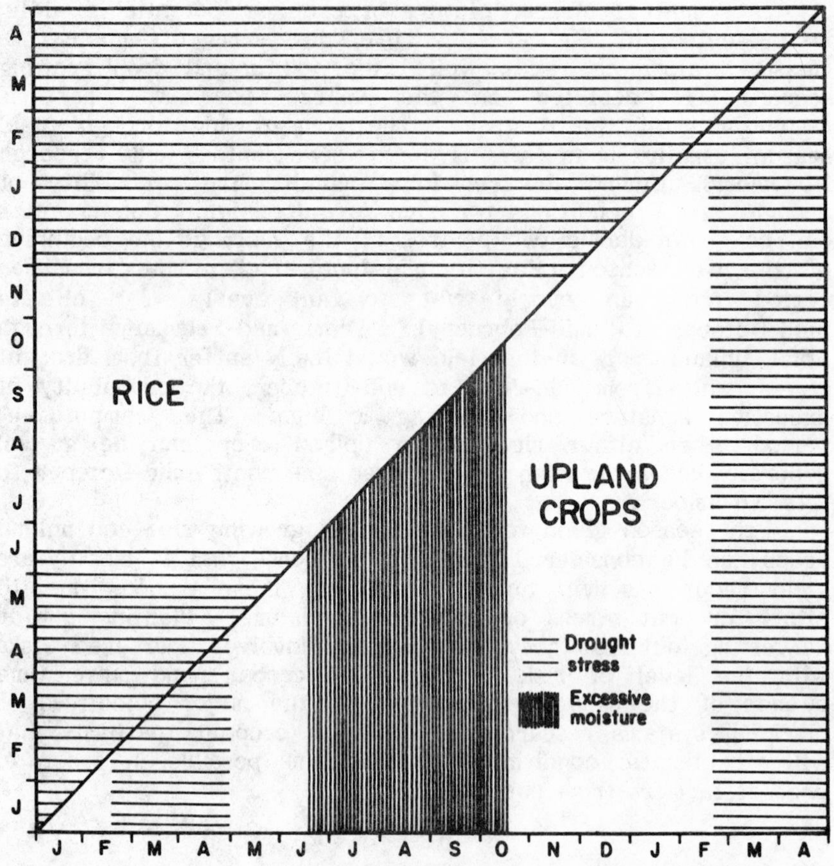

Figure 4.2 An agroclimatic map of cropping potential in the Iloilo rainfed area

Note: The environmental conditions for rice are specified above the diagonal; those for upland crops below the diagonal. These conditions correspond to areas with 5-6 wet (> 200mm) and 2-4 dry (<100mm) months per year

great deal of heterogeneity among farms even within a small geographic area, not all of the experiences described happened to all farmers, nor did they all happen to any one farmer. What is presented, then, is the description of a composite farm which illustrates the conditions and problems which small farmers face, and their typical reactions.

The family and their situation

Jose Umero lives with his wife Elena and their children in a small hut near their one hectare farm in a village in Iloilo Province. The Umeros are approximately 35, and have two sons who are too small to handle work in the fields.

The land that the Umeros farm is not irrigated, so they must depend upon rainfall for their crops. About one-half of their land is sloped, with light-textured soil, so that the soil quickly loses moisture when the rains stop even though the field is divided into level portions and bunded. The other half is flat land, just below the sloped portion, with heavier soil and better moisture retention capacity. Ordinarily, the flat land floods first because of its heavy-textured soil and because it also receives water which runs off the slope. The flat land, however, is not immune to water loss, as it too permits water to seep down to nearby fields which are lower.

The Umeros do not own the land they farm, but rent it on a harvest share basis. The landlord receives one-third of the output, net of harvesters' share, as payment for the use of the land. In a good year (high rice yield), the share of the crop the Umeros receive is enough to meet the family rice consumption needs, repay debts and provide seed for the next planting, with a little extra to sell in order to buy family consumption items. In a bad year (low rice yield), the Umeros cannot even harvest enough for their own consumption, and must rely upon relatives and friends who have a bit of surplus to lend them the items they need to subsist until the next harvest.

Since the Umeros began farming in 1963, their usual cropping system consisted of planting one rice crop followed by an upland crop or fallow. Until 1976, the rice Jose planted was of the traditional type; tall, late maturing and often plagued by diseases and insects. The varieties that gained greatest acceptance in the area were chosen as much for their eating quality as their yielding ability, since the yield of most varieties was much the same.

In the early 1970s the Umeros learned of a new type of rice, sometimes called "miracle rice", which did not grow as tall, was resistant to pests and diseases, and supposedly produced high yields. Jose was given some seed by the local extension agent. He was told that he must apply large amounts of fertilizer so the plant would grow and produce well, use herbicide to control weeds, and use insecticide to control even the insects that he could not see. Jose knew from the beginning that he could not afford to buy all the materials recommended by the extension agent. Nevertheless, he accepted the seed and chose to plant it in one of his better fields. He decided to treat the new rice just as he did the local varieties.

He sowed the new rice into a seed bed. The extension agent suggested that the miracle rice should be transplanted between 15 and 25 days after seeding. However, when that time arrived, the seedlings were too small and delicate. So, Jose waited until they were tall enough, about 40 days after seeding. Unfortunately, a typhoon blew the roof off his house. This delayed transplanting still further.

After transplanting, the Umeros kept a careful watch on the progress of their new rice, with anxious expectation of a bountiful harvest. From time to time while the crop was growing, Jose noticed tiny brown insects on the leaves of some of the plants. Surely these insects could not be causing trouble, since they were so small and did not appear to be doing much damage. He did, however, notice that some plants appeared to be much smaller than others. Jose thought that the seed the extension agent had given him had a mixture of at least two varieties. The seed people from the government were always doing that. That is why he preferred to save some of the grain from the previous harvest to use as seed for the next year's crop. Besides, it was cheaper and he did not have to go into debt to obtain seed.

What really worried Jose was the fact that the plants were not producing many tillers, which meant fewer panicles of grain to harvest. Perhaps, he thought, each panicle would be very long, which would make up for the reduced number. He thought about buying a little bag of fertilizer and applying it to see if things would change.

Jose wanted to sell two chickens to get one third of a bag of fertilizer, but Elena was against it. First of all, they could not yet be sure of a good harvest because as yet no grains were visible. Second, since the family had no cash on hand, the chickens were a ready source of cash to buy medicine in case one of the children fell ill, or food in case guests came, or for trading in case rice supplies ran low. She suggested that they wait until the grains were visible and filled, so they would be assured of at least harvesting something. Who knew whether a drought would come causing them to harvest nothing?

It was at times like these that Jose realized what a wonderful decision he had made when he married Elena. She seemed to have such good sense, such an ability to analyze, even though she had only barely finished primary school before she had to quit and go to work in the fields to help support her family. Jose felt disappointed about the fertilizer, but even more, frustrated because he could not provide a better life for Elena and the children. Nothing else to do that hot afternoon but collect some **tuba** (coconut wine) from the coconut trees, sell what he could, and sit out by the road in front of the local **sari-sari** store, watching the jeepneys and tricycles pass by while drinking away his worries.

When the panicles did appear and the grains were filling, Jose took the two chickens into town to sell in order to buy a small bag of fertilizer. When he came back, Elena seemed upset about something, but Jose wasted no time in applying all the fertilizer to the crop.

Because the rice matured before the wet season was over, the Umeros had to build a special shed in which to stock the rice in order to keep it from rotting in the field. Since the rice was wet when it was put in the shed, some of the grains germinated, while others rotted or turned brown. The wet rice was extremely difficult to thresh, so the Umeros had to spend a longer than normal amount of time in threshing. Winnowing was impossible until the rice dried out. Since daily rainfall was still common, the rice, placed on large straw mats in the sun, had to be carefully watched and hauled in and out of the shed as weather conditions changed.

When the rice was finally dried and winnowed, the yield was about three **cavans** (1 cavan = 44 to 50 kg), somewhat less than the Umeros normally harvested from that field using their traditional variety. Their expectations had not been fulfilled, even with all the extra expense and effort they had put in.

Jose felt that he had failed and went to see the extension agent who had given him the seed. He complained that the seed was mixed with other varieties, and that the shorter variety hardly produced at all. The extension agent asked Jose if he had followed all of the recommendations he had given him. Jose stated he had not, and began to explain the reasons why. While he spoke, the extension agent kept looking down and fumbling with some papers on his desk, making a comment or asking a question from time to time without looking up. When Jose finished speaking the agent look up, shrugged his shoulders and continued fumbling with papers on his disk.

On top of everything else, the milling and eating quality of the new rice was poor. When one of the rolling rice mills passed near their house, Elena rushed out with a bag of **palay** (unshelled rice) to try the new product. What she got back was disappointing, as the percentage of broken grains was quite high. She prepared the rice, but her family did not want to eat it. They did not like the taste, the texture was bad, and it wasn't sticky enough. Jose suggested selling it to the NGA (National Grains Authority) at the fixed government price, or to the Chinese middlemen who come around offering low prices for **palay** whenever the NGA warehouses were full and farmers are desperate to sell. Elena said she would be ashamed to sell it, and that she would feed it to the pig and use it to raise a few chickens.

Umero's cropping system since 1975

In 1975, Jose heard about experiments that were being conducted in the area on growing two crops of rice in one year. That sounded like a great idea if only it were possible. He knew of a local variety, Kapopoy, which farmers in the area usually interplanted with corn on their higher fields with lighter soils. They did this in years when a typhoon would bring rain before the normal onset of the wet season. The only thing that farmers could be sure about the rainfall is that it never followed the same pattern from year to year. The rains could begin early (April) and end early (September), they could begin late (July), and end early (September), or some intermediate pattern could occur. Furthermore, the period before the onset of the rains was often unstable; a typhoon, then drought, another typhoon, then drought again, and so on until the rains were no longer intermittent.

Jose dry seeded Kapopoy on his best field (most able to retain water) just after the first typhoon in April. Fortunately, heavy rains come in April that year and the Umeros were able to harvest a good crop of rice during early to mid-August. Shortly thereafter, Jose prepared a seedbed for the photoperiod sensitive variety, BE-3, which would mature in December. He then proceeded to prepare the land for transplanting, plowing twice and harrowing four times in order to puddle the soil and control weeds. By the end of September, all of the field was planted. The yield obtained at harvest in December was not as good as that of the previous crop, nor as good as yields they had obtained from BE-3 in previous years. Nevertheless, Jose and Elena decided that there was promise in double cropping and were glad to have the extra stock of rice. As it turned out, 1975 had been a very good year in terms of rainfall and the length of the growing season.

During 1975, the Umeros were contacted by people from the Cropping Systems Program of IRRI, and were asked to provide information about their daily farm activities, income and expenses, crop choices and a monthly inventory of their livestock. From the IRRI people, they heard about new rice varieties similar to the ones they had tried in earlier years. The Umeros were understandably skeptical, but liked the early maturation quality of the new varieties. Improvements, they were told, had also been make in eating quality and in pest and disease resistance.

Before the start of the 1976 crop season, Jose was able to obtain some seed for two new rice varieties being used by a neighboring farmer, IR-28 and IR-36. Even though farmers in the area had achieved good results with the new rice, Jose remembered his earlier experience with new rice varieties and was not willing to plant on his best land.

Largely due to their good luck the previous year, the Umeros decided to again direct seed (dry) Kapopoy on their lower field after the first April rain. He did the same with the IR-36 on a small parcel near the middle of this higher field. Rainfall during April after Jose had planted was scarce. The seeds germinated, and there was a fairly long spell without rain (two weeks). Many of the seedlings died, especially in the field where IR-36 was planted. Jose also noticed an unusually large number of weeds. There wasn't much he could do then, because he was busy preparing a seedbed and plowing and harrowing other fields which would soon be transplanted with the variety Kabangi. The dry seeded Kapopoy crop was also highly weed-infested, but a greater percentage of seedlings survived the drought because the lower field had retained more moisture.

The original stand in the IR-36 field was poor, so Jose broadcast the rest of the seed his neighbor had given him into the more sparsely populated areas. Weeds were thick and the crop continued to look bad, and Jose wondered if there would be any harvest at all. When he had time, he visited the fields of other farmers to compare their crops. Many of the farmers in the area had planted later than Jose, waiting until the fields were flooded before preparing land and broadcasting pregerminated seed (wet seeding). The stands in the wet seeded fields were much better than his, and weed problems were significantly less because farmers had had time to more thoroughly prepare their fields. The only thing Jose wondered about was whether or not it would still be possible to get in a second crop if one waited much after April to plant the first crop.

The Umeros' experience in 1976 with double rice cropping

did not turn out to be nearly as good as the year before. The dry seeded Kapopoy yield was down from the previous year. IR-36 and IR-28 matured at different times, so they were difficult to harvest; the field barely yielded enough for seed for the next planting season. Jose transplanted BE-3 as a second crop following Kapopoy, but the yield was low due to early termination of the rains. The Umeros worked hard during the dry season, greatly increasing the area planted to watermelon, in order to avoid falling too deeply in debt.

Many of the farmers who were able to plant only one crop of IR-36 produced more than Jose did with his double crop. Farmers seemed to prefer IR-36 over the other new varieties, both for its high yielding ability and for its eating quality. Many people found the flavor and consistency of IR-36 similar to that of the more popular local varieties.

Mainly due to the information gained the previous year, the Umeros decided to switch from dry seeding to wet seeding for the 1977 crop year. Jose's confidence in the new varieties was strengthened by what he had seen and heard, so he decided to plant nearly all of the farm to IR-36 and IR-28 with the seed that he had saved. Always in the back of his mind was obtaining a second rice crop, so he hoped for an early onset of the rains so he could get the first crop established as soon as possible.

Jose knew that his best chance for a second crop was on lower fields (plain), so he started land preparation in late May when heavy rains enabled puddling. He finished land preparation and seeded IR-36 at the end of June, and then began working on the sideslope. He finished wet seeding IR-28 on the sideslope by the middle of July.

The Umeros began to harvest IR-36 on the plain in the third week of September. For this process, Jose employed mostly hired laborers, who received 1/6 of the rice harvested for cutting, baling, hauling, threshing and winnowing. Elena supervised the measuring and took care of the drying. The IR-28 was ready for harvest a week later, so more hired laborers came in to harvest the sideslope fields.

Because the Umeros were hiring labor for harvest, Jose was free to begin land preparation for the succeeding crop. As the harvesters cleared a field, Jose began plowing it. His intention was to at least get the lower fields plowed and planted for a second rice crop. September rains were quite heavy, and land preparation on the plain was quite difficult due to heavy flooding. At the same time, Jose began to expect that heavy rains would continue or that the end of the season would be later than usual, as the beginning had been late. This meant that he might be able to get a good second crop on the sideslope if he could get it planted soon and rains continued. In the middle of October he decided to shift his land preparation activities to his higher sideslope fields. Jose worked extremely hard getting the land ready as soon as possible, because when the rains stopped, the higher fields would dry out quickly. He finished preparing and wet seeding all of his fields during the last week of October.

Unfortunately for the Umeros, the rains did not continue to be heavy, the rainfall levels dropping to less than 100 mm per month in November and December. The yields for the second rice crop were very low, with many of the fields producing little or nothing.

An Example of Notional Technology

Armed with a pretty thorough understanding of Iloilo rainfed farming systems in both agrocilimatic and socioeconomic aspects, it becomes possible to generate ideas about technologies that might significantly improve small farm productivity and income.

For example, the idea arose of trying to obtain increased food production through growing a **ratoon** crop. **Ratooning** of rice is the use of the plant's regenerative ability to produce a subsequent crop (or crops) from field stubble after the harvest of the first or planted crop. The expected yields of a **ratoon** crop are almost always lower than those of the main crop. The principal advantage, then, is the potential saving of both time and labor. The time-saving feature of **ratoon** cropping is what makes it most attractive as a potential new technology for rainfed areas with agroclimatic conditions similar to those prevalent in Iloilo.

The principal characteristics of a cropping pattern featuring **ratoon** are shown in Figure 4.3. Since the period from initial **ratoon** growth to grain maturity is short, the first rice crop can be established late enough so that the risk of drought stress at the beginning of the wet season is low. **Ratoon** growth begins immediately after (or sometimes before) the harvest of the plant crop. New shoots are produced at the base of the plant or grow from the nodes of previously cut tillers. Since **ratoon** matures in much shorter time than a planted crop, a rice-rice **ratoon** pattern can be undertaken within the limits of the rice growing season.

As Figure 4.3 indicates, during the latter third of the wet (rice growing) season, there occurs a period during which upland crops cannot be planted due to excessive moisture.

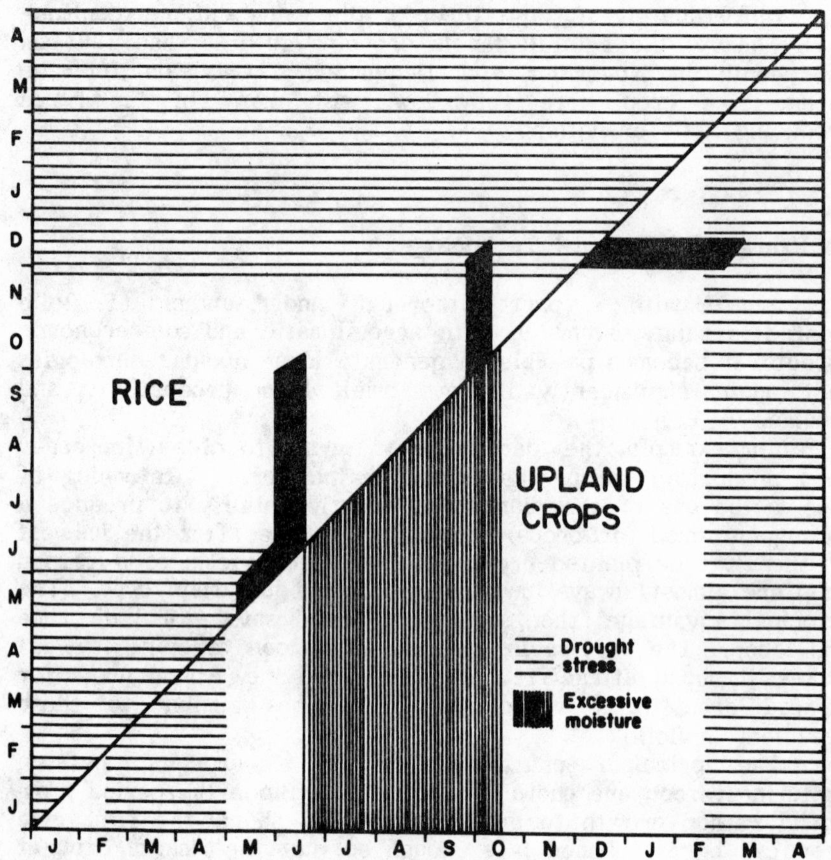

Figure 4.3 Ratoon rice cropping under Iloilo agroclimatic conditions

Furthermore, not enough time is left in the season to safely plant a second rice crop. **Ratoon** fits nicely in that "niche", making use of otherwise unproductive land. In many locations, the possibility would still exist for growing and harvesting an upland crop after **ratoon**.

Another feature that makes **ratoon** cropping potentially attractive in a rice-based cropping pattern is the reduced labor requirement. Since no land preparation or planting labor is required for **ratoon**, farmers may utilize idle land and still avail themselves of alternative employment opportunities such as harvesting on neighboring fields. **Ratoon** cropping then becomes a complementary rather than a competitive activity, as output can be generated with low levels of inputs and management.

Notional technology evaluation involves a variety of techniques for data collection and analysis. Farmer attitudes can be obtained through the use of specifically designed surveys which elicit farmers' perceptions of the applicability and desirability of the technology in question. Economic aspects of new technologies can be approximated through the use of mathematical modeling tools, such as linear programming. Both of these techniques were used in the present analysis, and their results are reported in detail in another publication (Chapman 1983).

Conclusions

The processes of identifying farm problems, and of deriving notional technologies for their solution, leads to two major conclusions that pertain to future activities in farming systems research. The first is that the generation of new technology appropriate to specific farming conditions should begin with a solid understanding of the needs and circumstances of those for whom the technology is designed.

The second conclusion reiterates the multidisciplinary nature of FSR. Since farmers' decisions and productive possibilities are highly affected by the agroclimatic and socioeconomic environments under which production takes place, the integration of information from both biological and social sciences will be necessary in order to achieve an understanding of the relevant system variables and the dynamics by which they operate. With such an understanding, the chances of developing new technologies which would alleviate farm problems will certainly increase.

The identification of notional technologies focuses on one step in the application of FSR methodology. At times this step is ignored or disparaged in agricultural research as "unproductive" or "background" information. Nevertheless, as demonstrated in the case described here, without a proper understanding of such background information agricultural research may easily be carried out which does not in fact resolve the problems for which it was commissioned. Given the expense of multi-year agricultural research efforts, careful analysis of notional technologies offers the possibility of substantial savings over the run of an agricultural research program, and improves the probability that the outcomes of the program will be used by farmers in the problematic environments where low levels of technology result in low productivity and incomes.

A number of technologies can be developed and tested in specific environmental situations. They may, however, be appropriate for use in other areas for reasons not yet understood by researchers. Though farming systems exist in a large number of environments, the number of technologies in some way suited to the different environments may be small. Technologies at research stations can be thought of as "seed" material. That is, the basic ideas and materials are developed and presented to farmers for **adaptation** before **adoption.** No single technological package will be wholly adopted by farmers, due to the variety of environmental conditions and limitations in different farming areas. Therefore, as certain parts of technological packages are accepted while others changed or rejected, farmers actually **invent** their own new technologies. There is no **one** appropriate technology for any area, so farmers should be provided with a number of technologies from which they can choose. A larger objective of research, then, is not developing one or more specific technologies; it is that of expanding the farmers' opportunity set. The farmers themselves are the ultimate judges of which technology, or set of technologies, is most appropriate.

A quote from Morss **et.al.** (1976) in an evaluation of the Plan Puebla project in Mexico summarizes well the major conclusion of this chapter;

> In the last analysis, the crux of the rural development dilemma lies less with persuading small farmers to adopt new behavior recommended by outsiders than it does with persuading outsiders to change their behavior and attitudes toward small farmers. And chief among the changes required of outsiders is the realization of their own vulnerability; that they do not have all the answers, that they cannot monopolize the process of rural development; that they cannot, in brief, help small farmers without the latter's assistance.

Notes

1. Agricultural Economist, CHEMONICS International Consulting Division, 2000 M Street NW, Suite 200, Washington DC, 20036. This chapter is based upon the author's Ph.D. dissertation (Chapman 1983) which was undertaken with the cooperation and support of the International Rice Research Institute, the US Agency for International Development, and the Agricultural Economics Department of Michigan State University. Responsibility for the material presented lies strictly with the author, and does not necessarily reflect the views or opinions of the institutions mentioned.
2. The development of notional technology described here took place under the sponsorship of the Cropping Systems Program (CSP) of IRRI.
3. Component technology involves changes in the management of single crops or crop mixtures which occupy a field during a single crop cycle.
4. An example of feedback from an economist to plant breeders is presented in Chapman (1979).

Bibliography

Anderson, J.R. and J.B. Hardaker. 1979.
"Economic Analysis in Design of Small Farmer Technology". In **Economics and the Design of Small Farmer Technology.** A. Valdes, G.M. Scobie and J.L. Dillon (eds). Ames, Iowa: Iowa State University Press.

Bradfield, R. 1972.
"Maximizing Production Through Multiple Cropping Systems Centered on Rice". In **Rice, Science and Man.** Los Banos, Philippines: IRRI.

Chapman, J. 1979.
"Ratoon Rice Culture as an Alternative Farm Technology for Rainfed Lowland Areas: An Iloilo Case". Paper presented at the Cropping Systems Program Seminar, May 29, 1979. Los Banos, Philippines: IRRI.

——. 1983. **Design and Analysis of Appropriate Technology for Small Farmers: Cropping Systems Research in the Philippines.** Ph.D. Dissertation. Department of Agricultural Economics, Michigan State University.

Cock, James H. 1979. "Biologists and Economists in Bongoland". In **Economics and the Design of Small Farmer Technology.** A. Valdes, G.M. Scobie and J.L. Dillon (eds). Ames, Iowa: Iowa State University Press.

Genesila, M., R. Servano and E.C. Price. 1979. "Socioeconomic Studies in Iloilo, 1976-78". Paper presented at the Annual Cropping Systems Workshop, March 1979. Los Banos, Philippines: IRRI.

Harwood, R. 1976. "Farmer-Oriented Research Aimed at Crop Intensification". In **Proceedings of the Cropping Systems Workshop.** Los Banos, Philippines: IRRI.

Hertford, Reed. 1979. "Biologists and Economists in Bongoland: Comment". In **Economics and the Design of Small Farmer Technology.** A. Valdes, G.M. Scobie and J.L. Dillon (eds). Ames, Iowa: Iowa State University Press.

Morss, E.R., J.K. Hatch, D.R. Mickelwait, and C.F. Sweet. 1976. **Strategies for Small Farmer Development.** Boulder, Colorado: Westview Press.

Pearse, A. 1980. **Seeds of Plenty, Seeds of Want.** New York: Oxford University Press.

Roxas, N.M. and M.P. Genesila. 1977. "Socio-Economic and Agronomic Characteristics of Existing Cropping Systems of a Rainfed Lowland Rice Area in Iloilo". Paper presented at the Annual Meeting of the Philippines Agricultural Economics Association. Cagayan de Oro City, March 12, 1977.

Zandstra, H.G. 1978. "Climatic Considerations for Area Based Rainfed Cropping Systems Research." Paper presented at a workshop on Management and Development of Rainfed Crop Production. Los Banos, Philippines: PCARR.

PART III

CASE STUDIES IN
ANNUAL CROP PRODUCTION

5

Farming Systems Research as an Alternative to Cropping Systems Research: An Example from Bangladesh[1]

Ben J. Wallace
and Ekramul Ahsan

Introduction

By the late 1960s, agricultural researchers had reason to be cautiously optimistic about their ability to ameliorate some of the problems of hunger and malnutrition confronted by many food deficit countries. New and improved food technologies, especially high-yielding grains, were by then available to these scientists. Agricultural researchers played a prominent role in adapting these new technologies to specific environments during the 1970s and many countries attained new heights of agricultural acheivement. These same researchers, however, continue to be faced with how to solve problems of hunger, malnutrition, unemployment and underemployment, poverty and public unrest in many countries (Sadikin 1983;3). And, despite the agricultural successes of the 1970s, there is some evidence to suggest that there may be at least one food crisis of global significance during the decade of the 1980s (see Ruttan 1982;18).

One of the important lessons learned from the experiences of the 1970s is that small farmers excel at managing scarce resources, scarce labor and limited capital. It has also become clear that the farmer must be placed at the center of any program aimed at rural economic development (Brady 1982;2). Some of these important findings grew out of the Cropping Systems Research (CSR) model, with its emphasis on cropping patterns (IRRI 1982, BARC 1981). The CSR model, however, despite its impact on agricultural researchers, administrators and extension agents is not adequate to fully meet the agricultural development demands of the 1980s. In this paper, we examine a CSR project in Bangladesh as illustrative of some of the problems of the CSR model and suggest how a Farming Systems Research (FSR) model, with its emphasis on the whole farm, has greater relevance for fulfilling the socioeconomic needs of the farmer. It is through FSR that the social scientist, especially the anthropologist or economist, potentially can make a lasting contribution to the agricultural and economic growth of many developing nations (see Norman, Gilbert and Winch 1979; Wallace 1984a; TAC 1978).

The Cost of Farming

Latif Rahman and his wife, Dulupa, are small farmers, possessing only one acre of land. On a plot located near their homestead, measuring sixteen decimals (.16 acre) in size, they planted jute during the spring season and mixed paddy during a later season. The land for each crop was plowed and harrowed three times, weeded three times, and harvested once. This work pattern is illustrative of most farmers in the Bangladesh village of Choto Kalampur.

Utilizing only his labor (his wife did not assist him) and two oxen, it took Latif Rahman one day to plow the sixteen decimal plot. He took another day to harrow the plot. He plowed the plot again five days later and then harrowed it again. In total, he dedicated six days to this activity. Because he had to rent the oxen, the cost of plowing and harrowing (excluding his labor) was taka 40/= per day. So, plowing and harrowing cost the household taka 240/= for each crop[2].

Jute seed was obtained by Rahman from the Bangladesh Jute Research Institute (BJRI) field station free of charge and the seeds for the mixed paddy came from his harvest the previous year. The household, then, had no cash expenses for the seeds for the two crops.

Rahman hired one day laborer to help him with planting the jute and the paddy. Planting each crop was accomplished in one day. He paid the laborer taka 15/= a day plus food.

The crops were weeded three times each during the growing season. Weeding, of course, is labor intensive work so it took six hired laborers (plus Rahman) one day to weed the sixteen decimal plot. They weeded each crop three times and were paid taka 15/= and food per day. For weeding the crop, then, the household had a cash outlay of taka 270/= plus food costs.

Harvesting the jute required two hired laborers for one day at taka 15/= plus food. The total cost for harvesting the jute was taka 30/= and food. Harvesting the mixed paddy required six laborers for one day at a cost of taka 15/= each plus food. Total cash cost for harvesting the mixed paddy was taka 90/=.

The total cash cost to the household for plowing, harrowing, weeding and harvesting the jute crop was taka 555/=. The cash cost for the mixed paddy was taka 615/=.

Processing the jute and paddy harvest was done by members of Latif Rahman's household. Importantly, however, most post-harvest activity is the responsibility of women in Bangladesh (See Wallace, Ahsan and Hussain 1985). Processing steps for jute in the region are; soaking the harvest in water, separating the fiber from the cane, cleaning the fiber, tying the cane into bundles, drying the fiber, storing and selling the fiber. The steps in mixed paddy processing are; husking, drying, wetting the paddy, parboiling, drying, milling and storing.

The cash cost per decimal of land farmed in the Choto Kalampur region is probably higher for a small farm household than for a large farm household. First, most large farmer households own their own oxen so they have only the feed and care cost for the needed animal power. Secondly, labor intensive work such as weeding is more cost efficient on larger plots of land. Because labor is hired by the day, workers tend to work at a speed that requires a full day to complete the job. Even if the laborers only work three-quarters of a day, they are still paid for the whole day. Six workers may take one day to weed sixteen decimals of land or the same six workers may take one day to weed twenty-five decimals of land.

In the case of Latif Rahman and Dulupa, it cost the household taka 555/= to produce a jute crop on sixteen decimals. A neighboring household planted thirty-two decimals of jute at a cash cost of taka 585/=. The neighbor, who owned his own oxen, used about the same number of person hours in producing the crop of jute as did Rahman. Importantly, however, the neighbor will get twice the jute crop yield as Rahman at about the same taka cost. This suggests that there is a plot size in this area of Bangladesh below which it becomes economically inefficient to plant certain crops.

Cropping Systems Research

In 1980 the BJRI established a CSR project in Choto Kalampur, a Bangladesh village of 351 households and 1800 persons. During the 1980 benchmark survey, the BJRI stratified village households, as follows; 1) Landless persons, 2) Landless tenants, 3) Small farmers owning .01 to 2.49 acres of land, 4) Medium farmers owning 2.5 to 5.0 acres of land, and 5) Large farmers owning more than 5.0 acres (BJRI 1981). In Bangladesh, over eighty percent of all farms are less than three acres and approximately fifty-four percent of the population is landless (IADS 1980). Most farmers in Bangladesh, then, are small farmers.

According to the BJRI field staff in Choto Kalampur, in order to be selected for the CSR project, the farmer had 1) to be "hard working", 2) to agree that there is "value" in experimental cropping trials, and 3) to own the "right" type of land for the trial. By the end of 1983, the BJRI field staff had persuaded only eighteen households in Choto Kalampur to participate in the on-farm trials, less than half of their goal of forty house-holds (BJRI 1982, 1983). In addition, all of the farm land utilized for the experimental plots was owned by either medium or large farmers. In brief, the CSR project of the BJRI was meeting with little success in Choto Kalampur.

Although the BJRI on-farm CSR project involves eight different experimental cropping patterns, for purposes of illustration only the pattern of jute-fallow-sweet potato will be examined here (BJRI Work Plan 1983-84). This cropping pattern involves four different experiments (Figure 5.1); 1. Fertilizer trials on jute (different combinations of nitrogen, phosphorus and potassium); 2. Component technology trials (different combinations of modern and traditional technologies); 3. Cropping pattern trials (BJRI controlled plots); and 4. Farmer's choice and method (a plot controlled and worked with traditional methods by the farmer).

Ideally, each fertilizer trial on jute and each component technology trial is conducted on a nine by nine meter plot of land. Between 500 and 1000 square meters of land are needed for the cropping pattern trials and, of course, land is also needed for the farmer's choice and method part of the experiment. Conducting this jute-fallow-sweet potato experimental pattern following recommended procedures would require a single plot of land approximately 6,500 square meters, or more than one acre. In Choto Kalampur, there are few individually owned plots of land measuring one acre in size (Wallace 1984a:30). Although the average land holding in Choto

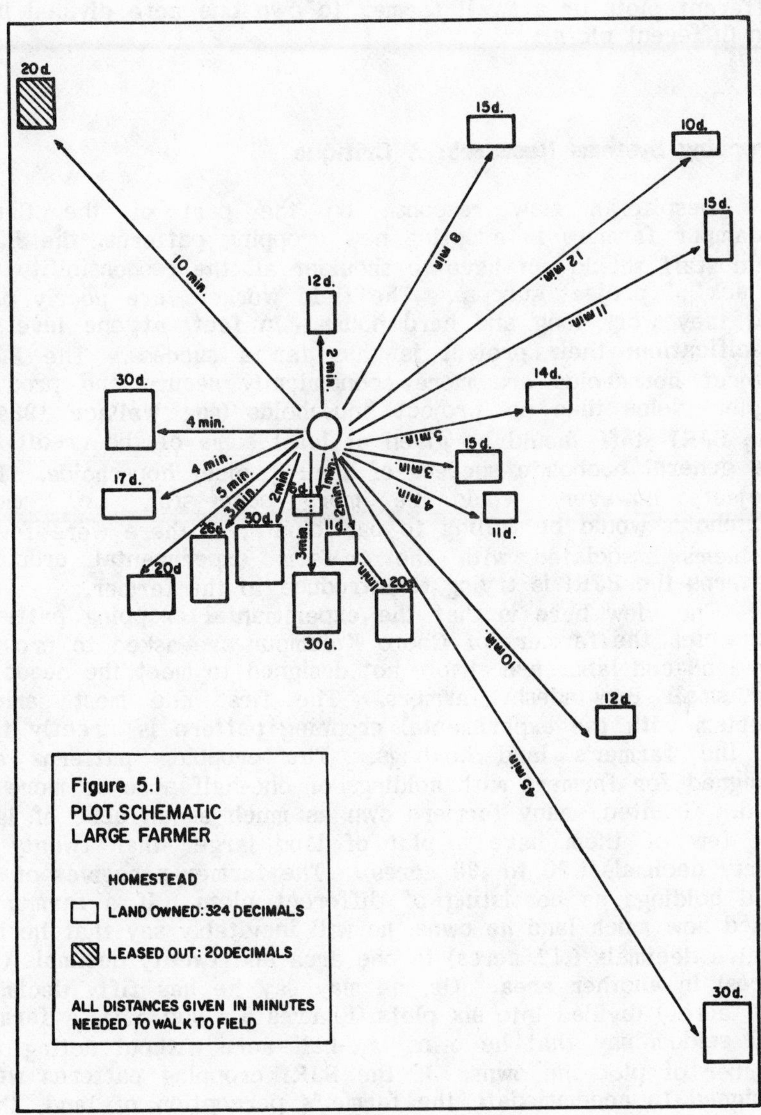

Figure 5.1
PLOT SCHEMATIC
LARGE FARMER

Kalampur is 2.06 acres or .85 hectares, a farmer's land is divided into many different plots, often ranging in size from .10 acres to .3 acres (Wallace 1984b). It is not unusual for a very large farmer to own four acres of land divided into twenty different plots or a small farmer to own one acre divided into ten different plots.

Cropping Systems Research: A Critique

Despite a slow response on the part of the Choto Kalampur farmers in adopting new cropping patterns, the BJRI field staff should not have to shoulder all the responsibility for a lack of project success. The field workers are poorly paid and they work long and hard hours. In fact, at one level of specification, their project is thus far a success. The BJRI project households are more economically secure and produce higher yields than non-project households (see Wallace 1984a). The BJRI staff should be given at least some of the credit for the general economic success of their project households. The project, however, would be more successful, i.e. more households would be willing to participate, if there were fewer problems associated with the on-farm experimental cropping patterns the BJRI is trying to introduce to the farmer.

The view here is that the experimental cropping patterns for which the farmers of Choto Kalampur are asked to provide the land and labor are simply not designed to meet the needs of the small Bangladeshi farmers. The first and most serious problem with the experimental cropping pattern is directly tied to the farmer's land holdings. The cropping patterns are designed for farmers with holdings of one-half acre or more of land. Granted, many farmers own as much as an acre of land but few of them have a plot of land larger than twenty to thirty decimals (.20 to .30 acres). The farmer perceives of his land holdings as consisting of different plots. If a farmer is asked how much land he owns, he will inevitably say that he has twelve decimals (.12 acres) in one area and twenty decimals (.20 acres) in another area. Or, he may say he has fifty decimals (.50 acres) divided into six plots (Figures 5.2 and 5.3). A farmer will seldom say that he owns one-half acre without noting the number of plots he owns. If the BJRI cropping patterns were designed to accommodate the farmer's perception of land, they would be more compatible with the farmer's world view and as a consequence, the farmer would be more likely to try them.

The second fundamental problem associated with the BJRI

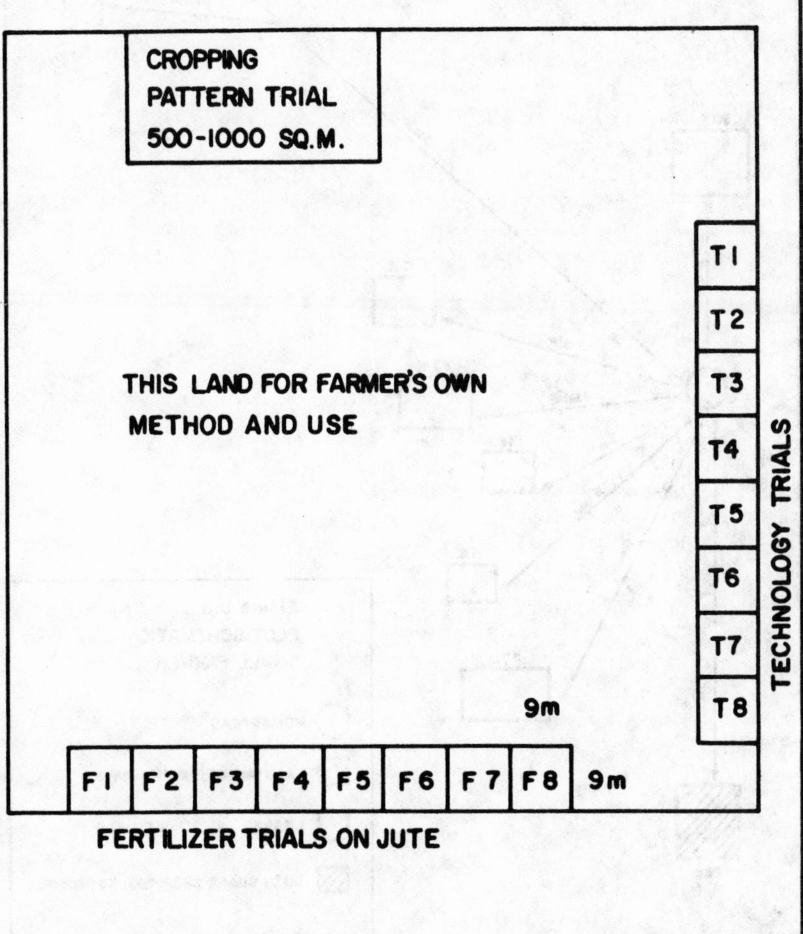

Figure 5.2
IDEALIZED EXPERIMENTAL PLOT ON HIGH RAINFED LAND USING JUTE

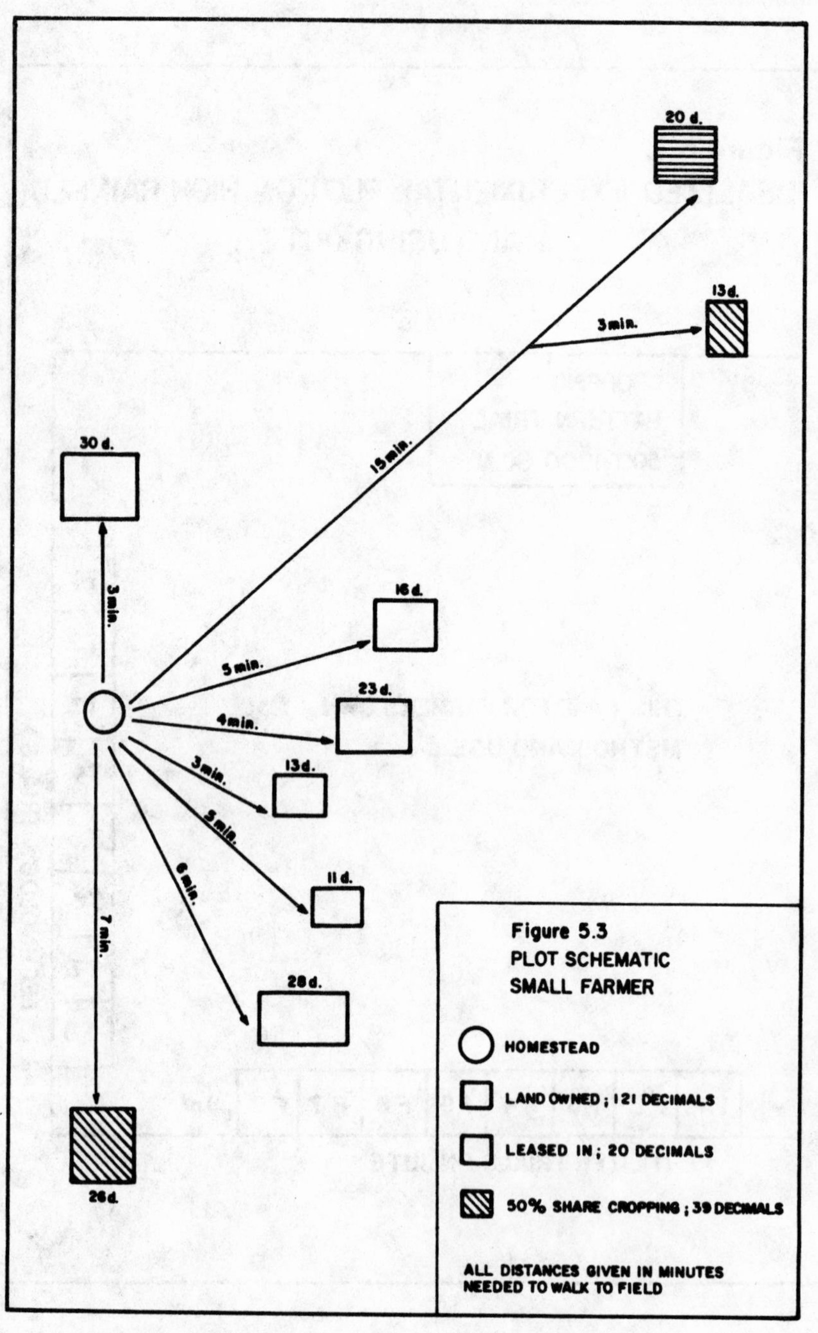

Figure 5.3
PLOT SCHEMATIC
SMALL FARMER

on-farm experimental cropping patterns is directly related to the farmer's perception of the cost and risk. With the possible exception of the large and medium land owners, farmers in Choto Kalampur live very close to economic disaster. A flood, a drought, or the loss of just one major crop can put a household into debt for many years. This fate, of course, is not new to the Bangladeshi farmer; it goes with being a farmer. But, new seeds, plowing techniques, seed drills, fertilizers and insecticides **are** new to the Bangladeshi farmer. Just because a large or medium farmer can use these new technologies does not mean that a small farmer can do the same thing. The larger farmer can afford to experiment. If he has a crop failure, his family will not starve nor will he have to go into debt to buy food. So, the poor farmer holds a "what if" attitude with regard to new technologies. "What if" the insecticides kill the crops? "What if" the new seeds won't germinate properly? The poor farmers know the risks involved with traditional farming methods. New methods suggest new risks to the poor farmer.

In addition to the risk problems, most farmers in Choto Kalampur have to consider the cost of improved technologies. If they don't believe that the cost is worth the risk even for a potentially higher yield, they are apt to stay with the traditional, but proven, methods. It is not that the farmers in Choto Kalampur are against new technologies and experimental cropping patterns. They are, for the most part, reluctant to venture into an unknown agricultural arena of which they know very little.

Farming Systems Research: Conclusions

Many of the problems associated with the BJRI Cropping Systems Research Project in Choto Kalampur can be rectified with the adoption of a Farming Systems Research approach. In the case described here, the CSR approach, despite its many strengths, especially in the arena of agricultural technology, places too little emphasis on the whole-farm. The FSR approach would have lead the BJRI researchers to the realization that their on-farm experimental cropping trials should have been simplified in order to obviate the fears of the farmers and redesigned to accommodate the limited and fragmented land holdings of the farmers.

The on-farm experimental trials designed for Choto Kalampur by the BJRI are best suited to the resources and risk taking capabilities of the larger land owners. It is not surprising then that most farmers have been reluctant to participate in the BJRI project because most farmers in Choto Kalampur are small land owners. Getting more small land owners to participate in the project is dependent on redesigning the experiments. The smaller land owner in Choto Kalampur, for example, owns around 1.5 acres of land divided into nine non-contiguous plots. It is unrealistic to ask a farmer of this type to devote a third of his land to an experiment because in his view, it may or may not be a good investment. He cannot afford to risk a third of his land on an experiment. Even if some of the farmers in the community are already getting higher yields from the experiment, he is apt to be hesitant. In his view, a guaranteed poor but predictable yield is better than an unpredictable high yield. On the other hand, the small land owner might be willing risk one small plot of his land on an experiment and the hope of a higher yield. This type of risk is not necessarily beyond his resources. If the experiment fails, he can still support his family from his other land and sources of income. If the experiment is successful, he will probably devote a little more of his land to the same cropping pattern the following year.

Bangladeshi farmers may be poor but they are hard working and have demonstrated an ability to rebound from both natural and political catastrophies. But, the future of the rural poor of Bangladesh as well as other parts of the world is only as good as the actions of the people responsible for planning, directing and implementing programs in directed social and economic change. One way for those of us involved with directed agricultural development is to work to ensure that the whole-farm and the farmer's perception of the whole-farm within a sociocultural and natural environment be given the highest research priority. The beneficiaries of this Farming Systems Research approach are scientists, administrators, extension workers, and most importantly, farmers.

Notes

1. The research on which this paper is based was conducted during 1983 and 1984. The research is a part of a long term Farming Systems Research project being conducted in collaboration with the office of Agricultural Economics and Social Sciences, Bangladesh Agricultural Research Council. Funds for the Research were provided by Winrock

International (formerly International Agricultural Development Service), Bangladesh Agricultural Research Council, and Southern Methodist University. We want to express our sincere appreciation to these institutions for their support. Finally, we also owe a debt of gratitude to the rural people of Bangladesh for opening their homes to us, for without their cooperation, this research never could have been completed. Although greatly indebted to the above individuals and institutions, we assume sole responsibility for the contents of this paper.

2. One U.S. dollar = 25.5 taka (1983).

Bibliography

BARC. 1981 **Cropping Systems Research Methodology.** Dhaka: Bangladesh Agricultural Research Council.

BJRI. 1981. **Socio-Economic Assessment of Improved Jute Production Technologies and Identification of Constraints to their Adoption.** Dhaka: Bangladesh Jute Research Institute.

——. 1982. **Socio-Economic Assessment of Improved Jute Production Technologies and Identification of Constraints to their Adoption.** Dhaka: Bangladesh Jute Research Institute.

——. 1983. **Socio-Economic Assessment of Improved Jute Production Technologies and Identification of Constraints to their Adoption.** Dhaka: Bangladesh Jute Research Institute.

——. 1983-84. **Bangladesh Jute Research Institute Work Plan, 1983-84.** Mimeo. Dhaka: Bangladesh Jute Research Institute.

Brady, N.C. 1982. "Opening Remarks". In **Report of a Workshop on Cropping Systems Research in Asia.** Los Banos, Philippines: International Rice Research Institute; 1-2.

IADS. 1980. **Report of the Review Team: Bangladesh Agricultural Research System.** Arlington, Va: IADS.

IRRI (International Rice Research Institute). 1982. **Report of a Workshop on Cropping Systems**

Research in Asia. Los Banos, Philippines. International Rice Research Institute.

Norman, D., D.H. Gilbert, and M.F. Winch. 1979. **Farming Systems Research for Agricultural Development: A Review of the Arts in Low Income Countries.** Washington, DC: USAID.

Ruttan, Vernon W. 1982. **Agricultural Research Policy.** Minneapolis, Minnesota: University of Minnesota Press.

Sadikin, S.W. 1982. "The Role of Cropping Systems in Increasing Food Production and Farmer Prosperity". In **Report of a Workshop on Cropping Systems Research in Asia.** Los Banos, Philippines: International Rice Research Institute; 3 - 10.

TAC. (Technical Advisory Committee). 1978. **Farming Systems Research at the International Agricultural Research Centers.** Washington, DC: Consultative Group for International Agricultural Research.

Wallace, Ben J. 1984a. **Working the Land: A Case Study in Applied Anthropology and Farming Systems Research in Bangladesh.** Dhaka: Bangladesh Agricultural Research Council.

————————. 1984b. "Diminishing Land Resources in Bangladesh". **South Asian Anthropologist.** 5:2; 103-108.

Wallace, B. J., R.M. Ahsan, and S. H. Hussain. 1985. **The Invisible Resource: Women's Work in Rural Bangladesh.** Dhaka: Bangladesh Agricultural Research Council, Technical Report.

6

Getting Marketing into Farming Systems Research: A Case Study from Western Sudan

Edward B. Reeves

Marketing is arguably the most neglected issue in farming systems research (FSR). While leading proponents of FSR frequently cite marketing as an important consideration (Collinson 1982; Gilbert **et.al.** 1980), none has attempted to explain in detail how the study of farmer marketing opportunities and constraints would actually contribute to specific FSR activities such as defining recommendation domains or improving the interpretation of on-farm agronomic trials[1]. Marketing does receive some attention in discussions of the diagnostic survey, or **sondeo**[2]. In this initial field reconnaissance and discovery phase of FSR, observation of local markets is necessary to make note of the availability of purchased inputs and to determine the farm products that are sold rather than consumed by the household. Useful as the checklist of "marketing factors affecting small farmers" (Shaner **et.al.** 1981:259-61) is for guiding interviews with farmers, agricultural extension agents, and local officials during a **sondeo**, it doesn't propose how specific marketing data should be collected and integrated into the later stages of FSR. I believe that there are two related reasons why no progress has been made to explicitly incorporate marketing analysis in FSR. The first is the overwhelming concern of FSR with alleviating biological constraints by the design, testing and recommendation to farmers of improved technologies. The second is a disciplinary bias which views marketing as an issue that is "exogenous" to the farm household and of concern largely in macro-level analyses that are carried out to inform policy-makers. Recent developments in the wake of the "Green Revolution" call into question the wisdom of these biases.

The contribution that economic anthropologists can make to FSR methodology lies in their emphasis on direct observation and behavioral analysis of marketing systems (Beals 1975;

Cancian 1972:77-95; Plattner 1983). To fully realize this, they must be familiar with the strengths of the macro-level analysis that agricultural economists do so well (e.g. Jones 1976; Mellor **et.al** 1968; Schmidt 1982; Southworth **et.al.** 1979). But, their particular contributions will lie not in the application of powerful, highly abstract theoretical models to the aggregate statistics of price movements and commodity flows[3]. Rather it will lie in concrete behavioral observations of farmer strategies for adapting to macro-level market conditions and the influence of this adaptation on farm management and the adoption of new technology. This means that the concept of farmer adaptive strategies (Bennett 1969, 1976; Barlett 1980), which has often been applied to farm production may also be applied to farmer participation in marketing. To illustrate how a behavioral study of farmer marketing strategies may contribute to biotechnical research of the sort that is advocated in FSR, I use the example of the well-known CIMMYT[4] method of economic analysis for agronomic research (Perrin **et.al.** 1976; Harrington 1982; Shaner **et.al.** 1981:125-38).

The objectives of the remainder of the paper are: (a) to explain the partial-budget analysis procedure advocated by CIMMYT (b) to critique the procedure by referring to a case study of farmer marketing strategies in Western Sudan, and (c) to draw conclusions about how a micro-level study of farmer marketing behavior can contribute to the CIMMYT procedure and to FSR in general.

Partial-Budget Analysis of Agronomic Trials

The partial-budget analysis that has been worked out by CIMMYT economists is designed to help agronomists select among alternative technologies, or treatments, the one that is most likely to be acceptable to a homogeneous category of farmers. Several assumptions are made: First, that the FSR team can distinguish a target group of farmers within an agro-climatic zone whose farms are similar and who have a similar form of management. Second, that having delineated this group of relatively homogeneous farmers, called a "recommendation domain" in FSR terminology (cf. Shaner **et. al.** 1982:215), it will be possible to conduct experiments under conditions representative of their farms. Normally, this requires the testing of new techniques and inputs on farmers' fields replicating their management capabilities and labor resources. Third, that alternative treatments used in the trials are not so complex and interactive that the effects cannot be measured with considerable accuracy. Fourth, that farmers think in terms of "net benefits" as they make farm decisions. This means that in evaluating a new technique or input the farmer will base his conclusion on weighing what he expects to gain from adopting the innovation against what he will have to give up. Finally, that the farmer's gains and costs, and by subtracting costs from gains, his net benefit, have money prices which can be estimated with sufficient accuracy to make the methodology worthwhile. When the actual local price of a farm input or product is known, that amount may be used to quantify the costs or gain. If the farmer's costs or gain cannot be measured directly (as in the case of family labor or food crops that are grown for family consumption rather than for sale), the concept of opportunity cost is employed to estimate a price for the input or product, since no money price is actually given up or received.

The CIMMYT methodology which is based on these assumptions is a form of marginal analysis referred to as partial budget analysis. It is designed to show, not profit or loss to the farm as a whole, but the net increase or decrease in farm income resulting from the proposed changes in techniques and inputs. Consequently, it involves only the gains and costs that vary with the treatments being compared. The objective of the analytical procedures is to derive recommendations from agronomic experiments that meet the requirements of maximizing net benefits to the farmer while avoiding the risk of unacceptable loss. An outline of the procedures is shown in the below:

Outline of procedures in Partial Budget Analysis

1. Calculate the average net benefits for each treatment.
    1.1. Estimate the benefits for each treatment.
        1.1.1. Calculate average yields for each treatment using customary agronomic procedures. Adjust yields first for differences between experimental management levels and farmer management levels. Then adjust for normal harvest and storage losses.
        1.1.2. Estimate the field price of the crop. This will be the local farmer market price less costs of harvesting, shelling/threshing, storage, and transportation from the field to the point of sale.
        1.1.3. Multiply field price times adjusted average yield for each product from the crop and sum to obtain gross field benefit for each treatment.

1.2. Estimate variable costs for each treatment.

1.2.1. Identify the variable inputs: those items which are affected by the choice of treatment. Estimate the quantity of each of these inputs used for each treatment.

1.2.2. Estimate the field price of each input. Normally this will be retail price plus transportation costs for purchased input. Field price of family labor will be an opportunity cost.

1.2.3. Multiply the field price of each input by the quantity and sum over inputs to obtain the variable cost for each treatment.

1.3. Subtract variable costs from gross field benefit to obtain the net benefit for each treatment.

2. Choose a recommended treatment using marginal analysis.

2.1. Array treatments from high to low in net returns. Eliminate dominated treatments. Calculate the rate of return to each treatment in capital. Graph the net returns curve if several treatments are involved[5].

2.2. Select as the recommendation the treatment which offers the highest net benefit and a marginal rate of return of at least 40% on the last increment of capital.

3. Check the suitability of the recommendation from the point of view of yield and price variability. The purpose of these additional analyses is to determine if the recommendation meets farmer standards for risk aversion.

3.1. Use minimum return analysis to compare the minimum returns from the selected treatment to those from all other treatments. If it compares unfavorably, a different recommendation may be more consistent with local farmer circumstances.

3.2. Use sensitivity analysis[6] to determine whether the choice of recommendation is sensitive to product or input prices which are particularly subject to estimation error. If the recommendation is sensitive to these changes, consider changing the recommendations or obtaining more information about the prices in question.

The contribution which an anthropologist could make to this sort of analysis is in the estimation of field prices and variable costs which accurately reflect the circumstances of different categories of farmers. My point is that farmer access to input and product markets can vary even when farming operations are homogeneous. One group of farmers may be relatively free to purchase or sell in the channel of their choosing and have partial control over their costs and the benefits received; other farmers may be constrained by their circumstances to choose less desirable options and consequently have less control. The prices found in the different market channels are linked with different farmer opportunities that may result in significantly different net benefits to the separate categories of farmers involved. To give substance to this argument I will review the circumstances of small farmers in the el-Obeid area of the Western Sudan[7].

Farmers and Marketing in the Western Sudan

Farming in the el-Obeid area is small-scale, family based, and commercialized to a considerable degree. Over 90% of the families produce food crops for sale and purchase much of the food that they eat. The rural population is dispersed in nucleated villages that vary in size from five or six households to 1,000 or more. Village population fluctuates seasonally. It is greatest during the rainy months when the crops are grown and least during the hot, dry season. The average household is composed of seven or eight members. Nuclear family residence, extended families, and other household arrangements are all common. Whereas the household is the basic unit of consumption (all members "eat from one pot"), agricultural production is typically managed by more than one decision-maker in the household. A typical pattern is for the husband and wife to manage separate farms. Unmarried sons and daughters who are old enough are also given land to cultivate and manage, when it is available. In addition to farming, virtually every household has members who work in off-farm occupations, which usually start after the harvest. The household head cultivates an average of 10 hectares, but the range of farm size is very great, from less than 2 to 30 hectares. While the outright sale of land is rare, one third of all cultivated lands are rented rather than owned by the farm operator. Most of the rented land is leased by better off farmers from farmers who are poorer than average. Labor is a key constraint. Therefore, farmers lacking the working capital to hire sufficient labor to cultivate all their arable land may still gain an income from the unutilized land by leasing it to someone else. The major field crops are sesame, millet, sorghum, groundnut, and roselle. Millet and sorghum are grown for household consumption, but surpluses are sold to other households and local middlemen. Millet grain is the preferred cereal staple of the rural diet, and the stalks of millet are used as a building material. Sorghum is frequently interplanted with sesame to stabilize the soil against wind storms. Sorghum grain like millet is used to make porridge; it is also used brewed into beer. Sorghum stover is a fodder. Sorghum is not nearly as important as millet in terms of the area devoted to its production or its role in the rural diet. Sesame is the principal market crop. About half of the cultivated area of most farms is planted in sesame. Most of it is sold to middlemen at various regional markets and is destined for processing into cooking oil that is consumed by the urban populations. Market prices of sesame tend to follow a predictable pattern. The market for groundnut, on the other hand, is quite unpredictable due to the volatility of the export market. groundnut planting varies with the previous year's market demand and farmer expectations of the current year's market. In the 1980-81 season, about 10 percent of the cultivated land was planted in groundnut after the previous season had established record high prices. Roselle is another crop whose cultivation is subject to volatile export prices. It is frequently intercropped with sesame and sorghum and is the least important of the major field crops.

The cropping cycle starts in April or earlier with land clearing. The field crops are planted between May and August. The uncertainty of rain leads to a strategy of phased planting of the crops and replanting patches where the crops fail to germinate. Farmers with more than 20 hectares to cultivate commonly hire labor for some operations. Laborers are compensated by cash wages that are negotiated between the farm manager and the laborer. The demand for labor varies unpredictably according to such factors as rainfall and rate of weed growth. Furthermore, the supply of free labor varies from village to village, again according to the sporadic distribution of rainfall as well as demographic factors, particularly the concentration of population in the vicinity of the village. The variation in wage rates between villages for a season and within the same village at different times in the same season can be substantial (Table 6.1). Moreover, whereas farmers with working capital can purchase labor for multiple weedings of their crops, farmers lacking the cash to purchase food for the family may neglect adequate weedings on their own fields in order to work on someone else's field and receive a wage. After the vagaries of climate, labor is the most important constraint on the cropping system. The cost of sustaining the household work force and of hiring additional laborers is the largest input expenditure. Capital investment by comparison is almost negligible since all agricultural operations are performed entirely with hand tools which are purchased at a modest cost from local suppliers and have a long use-life. The use of pesticides, particularly for groundnut, is beginning to come into practice; but as yet only a few farmers do this.

Generally speaking, farm management practices are homogeneous among all farmers in the region because of the hand-tool technology employed by all.

Transporting crops from the field to the village and after that to rural markets is accomplished by pack animals. A farmer who does not own a donkey to carry his crop can borrow (or rent) one from a neighbor. Transporting of crops from the villages to the central urban market at el-Obeid is accomplished by truck. A shortage of motor transport was not found in any of the villages studied. Villages which did not have a truck operated by one of their own inhabitants depended on regular visits from a transporter coming from a neighboring village. At the height of the marketing season, a village is usually visited by several trucks competing for business. Since crop marketing occurs in the dry season, the flooding of roads and tracks is not a problem. Long-term storage is also not a problem in this arid area because the highly reliable technique of pit storage is used.

Table 6.1
Wages* Paid for First Weeding of Millet (by area unit).

Village Name	Wage Rate	Village Name	Wage Rate
Umm Sot	5.000 - 10.000	Kazgeil	8.000
el-Kharta	5.000 - 15.000	Umm Arada	6.000 - 10.000
Demokia	8.000 - 15.000	Kaba	2.000 - 6.000
Umm Kuka	6.000 - 10.000	Umm Ramad	3.000 - 12.000
el-Hammadiya	6.000 - 12.000	Abu Haraz	10.000
el-Geifil	5.000 - 10.000	Wardass	10.000 - 14.000
el-Filia	4.000 - 10.000	Umm Sabagha	5.000 - 6.000
Burbur	7.000 - 10.000	Ayara	5.000 - 10.000
Bangedid	6.000 - 7.000	Bagbage	3.000 - 5.000

* Wages in Sudanese pounds. 1981 (1.000 L.S. = $ 0.90 US at official rate). All figures referring to Sudanese pounds are written with Sudanese notation, with a period rather than a comma marking the thousands column. The exchange rate noted above shows one thousand Sudanese pounds equivalent to 90 US cents.

The pervasiveness of the market in the lives of the farming peoples in the el-Obeid area can be gauged by the ubiquitous institution of the village shop. Local shopkeepers are responsible for the day-to-day provisioning of the farm household. Rarely are these individuals local monopolists. Shopkeepers depend on maintaining the goodwill of a clientele so that subtle forms of competition develop among them. In the larger villages the large number of shops leads to the development of periodic markets, which are usually associated with government-sponsored crop auctions that are open to all producers selling in lots of one sack or more. (Small farmers wanting to sell in smaller quantities may combine their separate marketings to make up an acceptable lot.)

A large wholesale crop auction is administered by government officials at el-Obeid. This market is for bulk sales so that in order to sell there small producers have to pool their products. Typically, the pooling is accomplished by villagers going together to hire a truck to haul a full load to the urban market. The auction market at el-Obeid was found to be the bellweather market for crop prices throughout the region. The prices posted at this market are usually lowest in the immediate post-harvest season (November), climb steadily through mid-season (January), and gradually decline after that. Both farmers and middlemen recognize this pattern as a recurring one. It has resulted in the characteristic marketing strategy which is elaborated below: farmers endeavor to hold the bulk of their crops off the market during the early part of the marketing season in order to take advantage of the rise in prices.

The rural and urban crop auctions are the formally instituted, official marketing channels. One purpose of the auctions is to permit the assessment and collection of taxes on the sale of the crops. This is a major revenue source for rural government. But there are other, informal channels which are important alternatives for farmers. Often, sale in these channels evades taxation; thus, they are officially frowned upon although the government lacks the police resources to prevent farmers and middlemen from taking advantage of them. These other channels include direct sales to el-Obeid merchants, sales to rural agents of the urban merchants, and sales to middlemen based in the farmer's own village. The latter are usually the owner-operators of shops and their crop-buying activities are part of the complex relationship these village merchants have with farm families. I turn now to an examination of that relationship.

The village shopkeeper has been maligned in many countries of Africa as a monopolist and exploiter of the rural population (Lele 1979), but at least one study in the Sudan argues that the farmer and the shopkeeper both benefit from their relationship (Wilmington 1955). An important step toward understanding the significance of marketing for the farming system in el-Obeid area is to comprehend how exchanges between farm families and village shopkeepers satisfy the consumption wants of the farm household. With more than half of the cultivated land producing crops for sale, the farm household is compelled to buy most of its food from local shopkeepers. Purchases are made daily or every few days, with the daily food bill ranging between 1.500 to 2.500 L.S.[8] for most families[9]. Thus, a central preoccupation of the household is to have money income to purchase food according to the daily requirements. A principal means to obtain the money income is from the sale of crops. But to sell immediately after the harvest entails a disadvantage because the prices are lower at that time than later in the marketing season. Farmers attempt to hold back as much of the crop as they can until the prices have increased. Still, they must have food, and money or a substitute for money is needed to buy it. This is where a major economic role of the village shopkeepers is engaged. Most shopkeepers will accept crops as payment for purchases of goods at the shop. The farmer adapts to this circumstance by selling only small amounts of crops to satisfy the immediate consumption needs, retaining the bulk for sale at a higher price.

Starting about midway through the dry season (February) and continuing until the time of the next year's harvest, farm families are confronted by another cash flow problem. Poor families and those with bad harvests run short in their stocks of crops to obtain a continuing money income or to exchange in kind for food. To some extent the money shortage is relieved by selling labor or taking up a seasonal trade. Nevertheless, many families experience a cash flow squeeze because household expenditures increase during the dry season as water becomes increasingly scarce and costly. As the cropping season begins with the onset of the rains, water is no longer expensive but now commercial sorghum, the mainstay of the diet when the millet stocks are exhausted, increases in price (100% in recent years) and there are the additional costs of farm inputs, particularly labor and sometimes seed. Again it is the village shopkeeper who provides a service that alleviates the strain. Goods from the shop are sold to farm families on credit. This is virtually a universal role of shopkeepers. Credit sales usually entail a promise by the purchaser to pay directly after the harvest in cash or in an amount of crops that has an equivalent money price. In the some instances the shopkeeper may put a service charge of 5-10% on the goods purchased. But according to farmers there is no extra charge in most cases if the customer is a relative or a close friend of the shopkeeper. Shopkeepers explain that credit sales allow them to move inventory at a season of the year when money is scarce. Offering credit is an important strategy for keeping a clientele, and foregoing the service fee amounts to a business promotion. When a credit balance is paid in crops the shopkeeper may gain, however; if the debt is paid early in the marketing season, the shopkeeper may store the crop for several months and receive a 20% increase in price.

While the cash flow problems of farmers are commonplace so that the relationship between shopkeepers and farm households just described obtains in a majority of the cases, it is not true that every farm family is hampered by a perpetual cycle of indebtedness. A sizeable minority of the farmers are able to avoid the cycle, and this has consequences that are seen in the different marketing strategies characteristic of indebted and debt-free farmers. The debt-free farmers have assets which can be liquidated when cash is required to meet household needs. Livestock are a major form of asset, and the larger crop producers regularly purchase animals with the surplus gained from crop sales. Owning a business is another form of asset. Over 90 percent of the shopkeepers were farmers. For them, shopkeeping is fundamentally an investment made by a successful farmer as a hedge against crop failure and agricultural market instability.

The general picture of the crop marketing system is that agricultural products move from the smaller villages to rural bulking centers and then to wholesalers, processors, and exporters in el-Obeid. The system appears to have a "dendritic structure" (Smith 1976) which would suggest that rural marketing centers are linked in highly stratified networks that sharply constrain the opportunities of all producers to market crops. Our study of crop marketing that used a sample of 166 farmers does not fully confirm this, however. It shows instead that a minority of farmers have greater flexibility in their marketing strategy than others. In comparing farmer access to channels it was found that larger producers are more likely to sell in only one channel while smaller producers sell in more than one channel (Table 6.2). This does not mean that the large producers have a more restricted opportunity. The opposite is the case. Field interviews determined the reason for this pattern is that smaller producers must opt for a diversified marketing strategy if they wish to maximize their returns. They have to sell small amounts of crops to local middlemen to purchase food while attempting to reserve the balance for later sale in a higher-return channel. Larger producers do not experience the cash flow constraint and are able to market all their crops in a single channel that brings the highest returns. They are also likely to store their crops until the prices are at the highest level expected for the season.

The patterns of marketing strategy become clearer if we

examine crop sales in the five channels (Table 6.3):

Urban Crop Auction: Twenty-three percent of the farmer sales transactions were at the crop auction in el-Obeid. Nearly half of all the producers' crops (by weight) are sold in this manner. On average, farmers market the largest quantity of crops in this channel (16 kantars) and earn the highest return of any legal marketing option (89-95%). The buyers in this channel are big merchants, warehousers, processors, and export agents who are licensed to bid in the auction. Taxes are assessed by the weight and auction price and are included in the sale price which the buyer pays. The producer pays for transportation and a 2% commission charged by the market agent who takes care of the crops during the weighing and auctioning procedures.

Table 6.2
Farmers' Mean Production (in mids*) of All Crops by Access to Channels.

Channel Access	N	Mean Production	Std. Error	Level of Significance**
only one	95	703.4	72.1	0.001
more than one	59	243.4	31.7	

* The "mid" is a Sudanese unit of volume equivalent to 4.125 liters.
** Student's t test for the difference of means.

Rural Crop Auction: Nineteen percent of producer sales transactions were in this channel, but the quantity sold amounted to only 7% of the total. This channel also had the lowest average quantity sold (3.2 kantars). Sales in the rural crop auction market are made to local village middlemen and occasionally to outside agents entering the market to compete with them. This channel appears to be the least important marketing alternative as far as the farmers are concerned. Generally speaking, farmers sell crops at the rural auction market when they intend to purchase expensive goods, such as livestock or clothing, at the periodic village market.

Table 6.3
Marketing Crops by Channel

	Channel				
	Urban Auction	Rural Auction	Shop Keeper	Agent	Urban Merchant
Proportion of farmer sales transactions	23%	19%	41%	9%	9%
Proportion of all crops (by weight)	43%	7%	29%	12%	10%
Average total quantity sold per farmer*	16.0	3.2	6.0	11.6	8.9
Estimated average return to the farmer by crop**					
sesame	95%	92%	88%	95%	98%***
groundnuts	89%	87%	75%	89%	91%***
roselle	97%	90%	83%	97%	97%***

* Expressed in "kantars". One kantar equals approximately 45 kg.
** The estimated percentage of the median daily price of crops at el-Obeid market which farmers may expect to receive by selling in each channel; these estimates are based on an informal survey of farmers, village merchants, and market clerks.
*** These estimates include a bonus which the urban merchant pays to the producer for conveying untaxed crops.

Village Shopkeeper: The highest proportion of crop sales transactions were in this channel (41%). The proportion of the total quantity of crops sold in this channel (29%) is second only to the urban crop auction. These results are consonant with the analysis above concerning the symbiotic relationship between farm households and shopkeepers. The returns to the farmer in this channel (83-88%) are the lowest of all five channels and this also explains why the average quantity sold is also rather low (6 kantars). Farmers said they would rather not market in this channel because of the low return, but it was convenient and sometimes necessary to do so when purchasing goods from a shop or paying off a debt. Unless the village shopkeeper reports these sales to a market clerk or sells the products at a crop auction, the crops sold in this channel are likely to escape taxation. However, survey results showed that, according to the shopkeepers' own reports, eighty-four percent of the crops sold were in a legal channel.

Agent: Nine percent of the producer sales transactions were made with rural agents working for the urban merchants, and this amounted to 12% of the crops by weight. Sales to agents brought producers the second highest returns (95-97%) and this is reflected in the high average quantity of crops sold (11.6 kantars). Direct sales by farmers to agents occur most frequently in the latter half of the marketing season (January through May), when farmers are opening storage pits and selling crops to pay for land clearing or to purchase commercial grain sorghum while the price is still low. This is a channel in which tax evasion occurs as a matter of course, since agents will not sell at the urban auction market unless they are apprehended by the police and compelled to do so.

Urban Merchant: Farmer sales transactions direct with the urban merchant account for 9% of all sales and 10% by weight of the crops sold. The average quantity of crops passing through this channel was 8.9 kantars while the estimated return to the producer was 91-98%, a little higher than selling to the rural agent. Marketing does not occur more often in this channel because of the risks to an independent transporter when hauling untaxed crops. Another reason that truck drivers mentioned was that lines of trucks back up in front of the merchant's warehouse. To avoid having the truck tied up for hours, the driver may go instead to the urban auction market where the load can be left with the commissioned market agent. The truck is then free to haul another load.

These data show that marketing alternatives are clearly available to producers. Forty-one percent of all sales transactions and 65% of all crops by weight were marketed in the higher-return channels (Urban Auction, Agent and Urban Market). The Rural Crops Auction and the Shopkeepers on the other hand — making up 60% of all sales transactions and 36 percent of all crops by weight — are relatively low return channels but they have a different function than the high-return channels. Transactions at local markets or with village shopkeepers occur in relatively small quantities when the strategy is to satisfy immediate consumption wants.

In their discussion of farmer access to markets Norman **et.al.** (1982:106-8) identify the proximity of developed markets and availability of transportation as major considerations in northern Nigeria. In the el-Obeid area, by contrast, transport services and market development did not pose very significant constraints. Student's t-tests for the differences in the mean quantities of crops sold between villages with a high level of institutional development (including the availability of transportation) and those with a low level of development reveals a significant difference in only the agent and urban merchant channels (Table 6.4). In developed villages there is a significantly higher average quantity of all crops sold to agents than in undeveloped villages. This finding is reversed in the case of sales directly to urban merchants. Since nearly the same proportion of sales transactions and quantities of crops sold occur with agents and urban merchants (Table 6.3), we may conclude that the farmers in the developed villages are more likely to sell to agents while in undeveloped villages the farmers are more likely to sell to urban merchants directly. This finding seems paradoxical. With their greater access to transportation farmers in the developed villages could be hypothesized to have greater opportunity to smuggle their crops directly rather than sell to an intermediary. But two circumstances discourage this pattern occurring. First, detection of smuggling is a greater risk in large developed villages where there are police officers and market officials charged with being on the lookout for smugglers. The risk of detection is less in undeveloped villages. Second, the agents in the developed villages are heavily capitalized compared to ordinary village shopkeepers who buy crops in exchange for goods. These agents have their own trucks and are practiced in smuggling and evading detection. The agent relies on the margin of profit that is gained from smuggling to offer farmers a relatively high price. In this perspective, agents and urban merchants seem to be alternative channels that are found in villages with different levels of commercial development.

The level of farm production, on the other hand, is a very significant indicator of farmer marketing strategies. The mean total production of the farmers selling in each channel was compared using a one-way analysis of variance (Table 6.5). A significant difference was found between mean production of farmers selling to the urban market and to village shopkeepers. This finding supports the analysis above concerning the relationship between cash flow and marketing strategy. Large producers are owners of non-crop assets such as livestock and business interests which provide liquidity. As a consequence they are able to market in a high return channel like that of selling to the urban market. Smaller producers facing cash flow constraints must market crops at the village shop.

Table 6.4
Mean Sales of All Crops by Marketing Channel and Village Institutional Development.

Village Environment	Channels									
	Urban Auction		Rural Auction		Shop Keeper		Agent		Urban Merchant	
	Mean Q*	N	Mean Q*	N	Mean Q*	N	Mean Q*	N	Mean Q*	N
Developed	668	11	234	18	293	31	812	8	70	3
Undeveloped	521	40	98	23	193	54	182	13	299	21
Significance	N.S.		N.S.		N.S.		0.052		0.003	

* Mean Q = Mean Quantity in "kantars". One kantar equals approximately 45 kg.

Table 6.5
Farmers' Choice of Channels by Mean Production (in mids*) of All Crops.

Channel	N**	Mean
Urban Auction	51	857.6*** (Shopkeeper)
Rural Auction	41	568.1
Shopkeeper	73	487.6*** (Urban Auction)
Agent	21	974.9
Urban Merchant	24	644.0
Total	210	629.8

ANOVA F ratio = 2.794. P Less Than .0273

* The "mid" is a Sudanese unit of volume equivalent to 4.125 liters.
** The samples are based on interview with 166 farmers. A third of the farmers marketed in more than one channel resulting in double counting. Since this would tend to make the means more similar rather than more different, the conclusions drawn in the text are made stronger by including farmers marketing in more than one channel.
*** Indicates significant differences between channel means by LSD Procedure (P Less Than .05).

Implications for the Use of Partial-Budget Analysis

I now turn back to partial-budget analysis and consider what the data from the western Sudan suggests about the methodology. My comments are addressed to three procedures: (a) estimating the field price of the crops, (b) estimating the field price of inputs, and (c) the identification of a recommendation domain — a homogeneous category of farmers. All three of these procedures are critical for the success of partial-budget analysis and the last one is also essential to any kind of FSR methodology.

To review, the field price of the crop is defined by the CIMMYT economists as: sale price less harvest costs, shelling/threshing, storage and transportation from the field to the point of sale.

The data presented above on farming systems and agricultural markets in the Western Sudan show that the field price of crops varies not only with the above factors but also with the marketing strategy which farmers are able to operationalize. Marketing strategy was found to vary with cash flow constraints experienced by the farm household and the level of total crop production. Larger producers, who do not experience a cash flow constraint, are able to reserve all their crops for sale when the prices are highest, and they are able to sell in a high-return channel. Smaller producers, experiencing chronic cash shortages, have to sell at least a portion of their crops at a disadvantageous time and in a low return channel. Moreover, the economist's practice of using average prices obtained from official market records to estimate the sale price would not reflect the prices received by most farmers in the el-Obeid area. Similarly, costs of inputs can be affected by marketing channel used. Consequently, access to beneficial prices and preferred marketing channels may be affected by farmers' economic conditions and production levels.

When accurate estimates of the field price of crops and variable inputs are not made, an accurate estimation of the net benefits of alternative treatments to farmers cannot be made, the partial-budget analysis is invalidated and doubt is cast on researcher recommendations. The result will discredit FSR and contribute to a heightening of the mutual misunderstanding and distrust that has been a dreary feature of relations between agricultural scientists and small-scale, limited resource farmers in many developing countries. In a worst case scenario, FSR in spite of its on-farm research approach could end up, like conventional agricultural research, giving the greatest assistance to those who least need it — the higher resource farmers.

This brings me to the issue of selecting recommendation domains of farmers. If FSR is to attain the goal of helping the world's poorer farmers it must be highly sensitive to all factors which differentiate farmers according to their advantages and disadvantages. The usual procedure in FSR is to differentiate recommendation domains according to production characteristics and directly related factors, such as climate, soil type, technology level, and general farm management practices. In the present paper I have tried to show that farmers who are highly similar on the bio-technical dimension may have different capabilities when it comes to product marketing and the management of cash flow to the household. This is a strong reason for including a behavioral analysis of farming marketing strategies in the delineation of recommendation domains. In the el-Obeid area, for example, I would make a distinction between larger producers, who market crops in high-return channels and are debt-free, and smaller producers, who market in low-return channels and are chronically indebted. Since a partial-budget analysis is pointless unless a valid recommendation has been identified, I would propose that a survey of farmer marketing behavior is an essential prerequisite to on-farm agronomic trials and their evaluation. I would be skeptical of a proposal to undertake on-farm trials immediately after a **sondeo** (Hildebrand and Waugh 1983) without a detailed survey of farmer circumstances including marketing opportunities and constraints.

Conclusion

I have criticized the methodology of FSR for neglecting the subject of farmer marketing strategies, and I have related this shortcoming to disciplinary biases in agricultural research and development. I have proposed that this omission opens an opportunity for economic anthropologists to participate in FSR projects. The study of farmer marketing strategies requires attention to regional marketing systems, particularly the channels for product and input marketing that are available to farmers, as well as to farm-level marketing constraints associated with the level of farm production and cash flow to the household. That an anthropologist's concern with understanding farmer marketing strategies can tie together both the micro-level, technological concerns of agronomists and the macro-level, institutional concerns of agricultural economists was demonstrated with reference to the partial-budget analysis developed by the CIMMYT Economics Program.

I conclude this paper with four additional remarks about the role of economic anthropologists in FSR, especially with reference to studying farmer marketing strategies. First, my emphasis here has been on the behavioral analysis of farmer marketing activities, but I do not regard this approach to be in competition with the cognitive approach to farmer decision-making, such as Gladwin (1982) advocates. The two approaches are complementary and both will contribute to FSR methodology. Second, I would stress, as Hildebrand and Waugh (1983) have done, that FSR is multidisciplinary research that is team-oriented. Anthropology has a highly productive tradition of individual field research, but the impulse "to do your own thing" must be eschewed when participating in an FSR project. Most of what anthropologists can contribute to FSR will be in helping the traditional agricultural disciplines to better grasp the realities of limited-resource farming. Anthropologists will have to learn to accept this supportive role without resentment and without succumbing to the urge to "chuck it". We are a discipline that prides itself on being sensitive to cultural differences. It is time that we developed some sensitivity toward the limits of our own disciplinary subculture and that we become willing to communicate with the members of other disciplinary subcultures as sympathetically as we do with the people of Bongo-Bongo. Third, anthropologists can take consolation in the fact that the task which FSR has set for itself is historical and immensely important for the welfare of much of the world population. If their role in FSR will be largely in support of the research of other disciplines, anthropologists should realize that understanding small farmers will tax all their conceptual and methodological skills. Finally, I can see nothing but benefits for the development of theory by anthropologists' participation in FSR. The issue of farmer marketing strategies alone would seem to call for the development of a new theoretical model that would integrate cultural ecology, regional analysis and political economy[10].

Notes

1. The major exception to this generalization, and it is only a partial one, is the work of Norman **et. al.** (1982). Even the marketing analysis in this study and its use of average margins for farmer sales begs the question whether variation in returns to different categories of farmers would significantly affect the economic evaluation of agronomic trials.
2. The **sondeo** is a preliminary reconnaissance of a region that is designed to discover broad characteristics of farming systems and major problems of concern to farmers. On the methodology of the **sondeo**, see Shaner **et.al.** (1982:289-93).
3. Anthropologists have been in the forefront of criticizing overly abstract and rationalistic, economic models of farmer decision-making (Barlett 1980; Cancian 1972; Ortiz 1983). For a critique of macro-economic analyses in the Sudano-Sahelian zone of West Africa see Harriss (1982).
4. Centro Internacional de Mejoramiento de Maiz y Trigo.
5. A "dominated treatment" has lower net benefits **and** higher variable costs than some other treatments in the experiment. Dominated treatments are eliminated from further consideration because they are clearly inferior to some other treatments. The purpose of calculating the rate of return to each increment of capital is to determine if it meets farmer minimum requirements for investment. The CIMMYT economists believe that a 40% return is generally the minimum rate that small farmers will accept (Harrington 1982:17-23).

6. Sensitivity analysis determines the effect of price variation on a treatment that is being considered for recommendation to farmers. Very simply, it entails calculating the marginal return (see above) **twice**, once using a high but likely price and once using a low but also likely price (Harrington 1982: 35-37).
7. The information was gathered during a 14 month study (1981-82) of small farmers and rural markets in the el-Obeid area of North Kordofan, Sudan. The study combined in-depth interviews with structured surveys in 15 villages. A close collegial relationship was secured with agricultural scientists working in the region. The project was sponsored by the International Sorghum-Millet Research Project and the Departments of Anthropology and Sociology at the University of Kentucky, with funding provided by the U.S. Agency for International Development.
8. In 1982, 1.000 L.S.=$US 1.11.
9. A survey revealed that farm families produced on the average only enough millet and sorghum grain to supply the household for about six months. For the remainder of the year commercial sorghum is purchased from local suppliers.
10. For a survey of the literature on cultural ecology that would be of interest to FSR practitioners, see Barlett (1980). Smith (1976; 1983) is a key resource for regional analysis. On the political economy perspective, see Nash (1981), Wallerstein (1979), Wolf (1982) and Worsley (1984).

Bibliography

Barlett, Peggy F. 1980.
"Adaptive Strategies in Peasant Agriculture." In **Annual Review of Anthropology** 9; 545-73.

———— (ed.) 1980.
Agricultural Decision Making: Anthropological Contributions to Rural Development. New York: Academic Press.

Beals, Ralph L. 1975.
The Peasant Marketing System of Oaxaca, Mexico. Berkeley: University of California Press.

Bennett, John W. 1969.
Northern Plainsmen: Adaptive Strategy and Agrarian Life. Arlington Heights, Illinois: AHM.

———— (ed.) 1976.

The Ecological Transition: Cultural Anthropology and Human Adaptation. Chicago: Aldine.

Bennett, John and Don Kanel. 1983.
"Agricultural Economics and Economic Anthropology: Confrontation and Accommodation." In **Economic Anthropology: Topics and Theories.** Sutti Ortiz (ed.). Lanham, Maryland: University Press of America; 201-47.

Cancian, Frank. 1972.
Change and Uncertainty in a Peasant Economy. Stanford: Stanford University Press.

Collinson, M.P. 1982.
"Farming Systems Research in Eastern Africa: The Experience of CIMMYT and some National Agricultural Research Services, 1976-1981." **MSU International Development Paper, No. 3.** East Lansing, Michigan: Michigan State University, Department of Agricultural Economics.

Epstein, T. Scarlett. 1975.
"The Ideal Marriage between the Economist's Macro Approach and the Social Anthropologist's Micro Approach to Development Studies." **Economic Development and Culture Change.** 24(1);29-45.

Gilbert, E.H., D.W. Norman, and F.E. Winch. 1980.
"Farming Systems Research: A Critical Appraisal." **Rural Development Paper No. 6.** East Lansing, Michigan: Michigan State University, Department of Agricultural Economics.

Gladwin, Christina H. 1982.
"The Role of a Cognitive Anthropologist in a Farming Systems Program that has Everything." In **The Role of Anthropologists and Other Social Scientist in Interdisciplinary Teams Developing Improved Food Production Technology.** Los Banos, Philippines: International Rice Research Institute.

Harrington, Larry. 1982.
"Exercises in the Economic Analysis of Agronomic Data." **Economic Program Working Paper.** Mexico City: CIMMYT.

Harriss, Barbara. 1982.
"The Marketing of Foodgrains in the West African Sudano-Sahelian States: An Interpretive Review of the Literature." **Economics Program Progress Report No. 31.** Andhra Pradesh, India:

ICRISAT (International Crops Research Institute of the Semi-Arid Tropics).

Hildebrand, P.E. and R.K. Waugh. 1983.
"Farming Systems Research and Development." **Farming Systems Support Project Newsletter** 1(1);4-5.

Jones, William O. 1976.
"Some Economic Dimensions in Agricultural Marketing Research". In **Regional Analysis** Vol. 1. Carol Smith (ed.) New York: Academic Press; 303-26.

Lele, Uma J. 1979.
The Design of Rural Development: Lessons from Africa. Baltimore: Johns Hopkins University Press.

Mellor, John W., T. F. Weaver, U.J. Lele and S.R. Simon. 1968.
Developing Rural India. Ithaca: Cornell University Press.

Nash, June. 1981.
"Ethnographic Aspects of the World Capitalist System." **Annual Review of Anthropology.** 10;393-423.

Norman, David W. 1982.
"The Farming Systems approach to research." **Farming Systems Research Paper No. 3.** Manhattan, Kansas: Kansas State University.

Norman, D.W., E.B. Simmons, and H.M. Hays. 1982.
Farming Systems in the Nigerian Savannah: Research and Strategies for Development. Boulder, Colorado: Westview Press.

Ortiz, Sutti. 1983.
"What is Decision Analysis About? The Problems of Formal Representations." In **Economic Anthropology.** S. Ortiz (ed.) Lanham, Maryland: University Press of America; 249-297.

Perrin, R.K., D.L. Winkelmann, E.R. Moscardi and J.R. Anderson. 1976.
From Agronomic Data to Farmer Recommendations: An Economics Training Manual. Mexico City: CIMMYT.

Plattner, Stuart. 1983.
"Economic Custom in a Competitive Marketplace." **American Anthropologist** 85(4); 848-58.

Schmidt, Gunter. 1982.
The Efficiency of the Marketing System for Agricultural Products in Pakistan's Punjab. Fort

Lauderdale, Florida: Breitenbach Publishers.
Shaner, W.W., P.F. Philipp, and W.R. Schmehl. 1982.
Farming Systems Research and Development: Guidelines for Developing Countries. Boulder, Colorado: Westview Press.
Smith, Carol A. 1983.
"Regional Analysis in a World-System Perspective: A Critique of Three Structural Theories of Uneven Development." In **Economic Anthropology**. S. Ortiz (ed.) Lanham, Maryland: University Press of America; 307-59.
———— (ed.) 1976.
Regional Analysis: Economic Systems. Vol. 1. New York: Academic Press.
Southworth, V.R., W.O. Jones and S.R. Pearson. 1979.
"Food Crop Marketing in Atebubu District, Ghana." **Food Research Institute Studies** 17(2); 157-95.
Wallerstein, Immanuel. 1979.
The Capitalist World-Economy. Cambridge: Cambridge University Press.
Wilmington, Martin W. 1955.
"Aspects of Money Lending in Northern Sudan." **Middle East Journal** 9(2); 139-46.
Wolf, Eric R. 1982.
Europe and the People Without History. Berkeley: University of California Press.
Worsley, Peter. 1984.
The Three Worlds: Culture and World Development. Chicago: University of Chicago Press.

7

Trials and Errors: Using Farming Systems Research in Agricultural Programs for Women

Anita Spring

Introduction

In many African countries, extension programs for women consist of home economics courses that mostly focus on training in "stitching" (sewing) and "stirring" (cooking). Often the women for whom these programs are designed are rural farm women engaged in agricultural production activities as well as in domestic and reproductive activities. Because of the emphasis on home economics, and particularly on a narrow definition of home economics, extension programs for these women farmers do not provide instruction on cultivating crops, caring for livestock or managing farms. Additionally, this type of training does not provide the credit and inputs that will help these women assure food security and income for their families. The notion of scientific agriculture for men and scientific home economics for women is part of a model used in the land grant universities in the United States and in other institutions in developed societies. In terms of agricultural production, an assumption of the model is that women are considered as helpmates on the farm or as farmers' wives who may be interested in raising poultry and cultivating small vegetable gardens for household consumption, but who are not considered as farmers in their own right. The model has been transferred to many developing countries causing women's programs to focus on domestic and on reproductive aspects of women's roles and to ignore the productive aspects of women's roles (Mead 1973; Gladwin and Staudt 1983a, 1983b; Gladwin, Staudt and McMillan 1984).

This paper examines programs for rural women in Malawi

that until recently emphasized home economics, particularly cooking and sewing courses, and did little to provide instruction in agricultural production. It examines how an FSR approach that included farmer-managed demonstrations and trials was used to include women farmers in agricultural programs. The paper describes a project that enabled extension agents and researchers to work with a group of farmers often by-passed and to re-orient training for women to include agricultural training. Furthermore, the argument presented here attempts to distinguish between problems that affect smallholders in general and problems that are gender specific, and in particular which affect women agriculturalists.

The Extension Situation in Malawi

The Republic of Malawi is a landlocked country with 6.5 million people located in south central Africa. The economy is based on agricultural production that is divided between private and government estates and smallholder farms. The state sector contributes approximately 70% to the agricultural exports whereas the smallholder sector contributes 30% to export in addition to feeding itself. In 1977 the country embarked on a 20 year National Rural Development Plan to assist farmers in the smallholder sector that the government felt was lagging behind the estate sector (NRDP 1977). To accomplish these aims, the country was divided into eight contiguous units called Agricultural Development Divisions (ADDs) administering development projects. Each ADD administered two to six Rural Development Projects (RDPs). The administrative staff at ADD and RDP levels supervised the extension staff who were involved in delivering services such as training, credit and inputs to farmers.

Traditionally, women were heavily involved in Malawi's smallholder farming sector, doing 50% to 70% of the farming operations (Clark 1975; Spring, Smith and Kayuni 1983b). Recent studies show that women are increasing their agricultural production activities and are becoming the full time farmers on the family land as men become part-time farmers, spending less time on their farms because of off-farm wage activities (Kydd 1982; Spring, Smith and Kayuni 1983b; Spring 1984). Although women are heavily involved in agricultural production, until recently, women's extension programs organized by the Ministry of Agriculture (MOA) focused almost exclusively on home economics. The MOA has a large extension staff of about 1,950, approximately 1,800 of whom are men and 150 are women. The male extension workers were trained in agricultural subjects so that they can deal with rural men and the female staff were trained in home economics subjects to deal with rural women. Seventy-five percent of the male extension worker's two year training was in agriculture, 78% of the female extension worker's 6 months to one year training was in home economics (Spring 1983). Until recently most of the agricultural training for women extensionists focused on poultry raising and vegetable production and they in turn transferred this information to rural women. Because of the differential training mandate for male and female extension workers, courses for rural people were differentiated by sex as well. Table 7.1 shows that content of training courses for rural people at day and residential centers consisted of 75% to 84% home economics for rural women and 88% to 93% agriculture for rural men. The remaining 16% to 25% of women's training consisted of courses in health, leadership, poultry raising and vegetable production. The remaining 7% to 12% of training for men was in health and leadership (Spring 1983).

In 1981 the MOA changed the designation of the Home Economics Section to Women's Programs with the hope of increasing agricultural services to rural women. It was during this change that the Women in Agricultural Development Project (WIADP), funded by the Office of Women in Development of USAID, operated in Malawi (Spring 1985). The WIADP was administered and managed by an anthropologist (the author of this paper) and staffed by an American agronomist, a Malawian general agriculturalist and Women's Program's Officer, and an American agricultural economist, the latter for only part of the Project's life. The WIADP was under the auspices of the MOA and reported to the Departments of Research and Agricultural Development (Extension). Staff members of these departments assisted the WIADP throughout its duration.

The WIADP was conducted between 1981 and 1983 and aimed to document women's and men's involvement in smallholder agriculture using a FSR approach, to ascertain problems facing women farmers as client groups, to assist the Women's Programs Section to reorient its direction from home economics to agriculture, and to work with farmers, extension agents and research personnel to develop workable communications patterns and solutions to problems. To achieve these ends, a variety of research and training activities were conducted including the use of the FSR methodology in different parts of the country. Demonstrations and trials on soyabeans were conducted in one area by the WIADP (those discussed here), and the WIADP assisted in **sondeos** (or rapid reconnaissance surveys) and intercropping trials (of maize, cowpeas and sunflower) that were

Table 7.1
Breakdown of Classroom Time for Male and Female Farmers Taking Agricultural and Home Economics Courses at Day Training Centers, Residential Training Centers and Farm Institutes (percentages).

Subject Matter	Day Training Centers		Residential Training Centers		Farm Institutes	
	Male (Agric)	Female (Home Ec)	Male (Agric)	Female (Home Ec)	Male (Agric)	Female (Home Ec)
Crops	48	19	44	16	33	8
Livestock	13	6	20	6	50	8
Farm Management	27	0	22	0	10	0
Total Agriculture	88	25	86	22	93	16
Health & Nutrition	12	45	6	49	3	52
Clothing	0	30	0	23	0	27
Leadership	0	0	8	5	4	5
Total Non-Agricultural	12	75	14	78	7	84

Source: "Syllabus for Farmer Training Centres of the Department of Agriculture Development," Lilongwe, Malawi: Ministry of Agriculture, n.d.

conducted by the Farming Systems Analysis Section of another USAID funded project (see Hansen in this volume). This paper focuses on demonstrations with soyabeans conducted in 1981-82 and trials in 1982-83. A **sondeo** had already been carried out (Hansen 1981; Hansen and Ndengu 1983) in the Lilongwe Rural Development Project (LRDP), the area under consideration.

The LRDP, one of the first development projects in Malawi, was begun in 1968 under funding from the World Bank. The LRDP is one of five projects in the Lilongwe Agricultural Development Division (LADD) located in Central Region. The LRDP is located to the west and south of Lilongwe city, the capital, in an area of gently undulating plains. The area has an altitude of 1,090 to 1,230 meters, a temperature range of 15C to 23C, and an average annual rainfall of 640 mm to 1,090 mm concentrated between November and April. The soils are moderately fertile and well suited for growing maize, groundnuts and tobacco. The area was originally selected for funding in the late 1960's because it was in the major granary area of Malawi, the people were accustomed to cash cropping, and the farmers had shown interest in land reorganization and improved farming (Kinsey 1973; Lele 1975).

The structure of LRDP is set up such that six administrative groups oversee 40 units (recently renamed Extension Planning Areas, EPAs) that have been developed since 1968. Each unit has a Development Officer and a number of "grass roots" technical assistants, typically including four or more general extension agents and occasionally other personnel such as tobacco, forestry, and livestock/veterinary assistants who are male; about half the units have one female extension agent. During the first two years of LRDP, there was one extension worker for every 200 families compared to one for every 1,200 to 1,300 families in the non-project area. In addition to the extension staff, there are development/planning committees for farmers organized at village, unit, group and project levels. Day training courses take place at the unit (or EPA) center while short courses are given at the two residential training facilities. Most agricultural courses are for men while women receive home economics courses. Needless to say, because of the ratio of extension agents to rural people of both sexes, many men and women are unable to receive any contact at all (Spring 1984).

People in the LRDP constitute a mix of farmers: those at subsistence levels, those who obtain varying amounts of income from agriculture, and those who seek wage labor in nearby Lilongwe or who work on agricultural estates in other parts of the country. People primarily monocrop maize, groundnuts, tobacco, beans and sweet potatoes under rainfed conditions with the average landholding being about 4 acres per household (Kinsey 1973; Lele 1975; NSO 1982). Kydd (1982) estimated that in 1968-69 nine percent of the men were away from the area and that 11% of the households were headed by women. By 1978-79, his figures showed that 28% of the households were headed by women and that men without sufficient land tended to migrate. The 1980-81 National Sample Survey of Agriculture carried out by the Ministry of Agriculture showed that 20% of the households were headed by women (NSO 1982) with 39% of these female household heads being married to men who were away from the farm (Spring 1984).

Soyabeans: A New Commodity for Smallholders

For many years, soyabean had been grown commercially in the estate sector for the bean and as a green manure. Its production in the smallholder sector was negligible. In 1981, the Women's Programs and the Food and Nutrition Sections of the MOA introduced the preparation of soyabeans to women extension agents who were attending their annual national refresher course. The Food and Nutrition and Women's Programs Sections were placing emphasis on the crop because the Malawian diet is particularly deficient in fats and to a lesser degree deficient in proteins; soyabean contains about 40% fat and 20% protein. Women extension agents were taught recipes for preparing soyabean milk, porridge, coffee, snacks, relish, scones and other baked goods (Spring 1981). Several months later, the WIADP, in the course of investigating farmer production and training, came across a woman extension agent who had attended the refresher course and who subsequently introduced the soyabean recipes to her home economics class of sixty-four women farmers. However, since she herself had not received any information on the cultivation of soyabean or on inoculation (with **Rhizobium** bacteria — see below), she was unable to teach the women how to grow the crop properly. She was planning to give the women a handful of seed so that they could grow the crop as best they could.

(The standard extension activities for women in Malawi usually focus on a craft, sewing or cooking activity; sometimes these activities are thought to be for "income generation". They may in fact function to preserve home income, since markets for these products are few and these types of activities rarely net any sizeable income for the participants. Basically the situation encountered in LRDP was an example of the way in which courses for women operate. Recipes using new and sometime exotic materials often are taught to trainers who introduce them to rural women. The women may be unable to sustain the technique because of lack of materials. In addition, large groups of women must make do with small amounts of materials. Technical information may be lacking. By contrast, training for male farmers tends to start with inputs that are available, with markets to sell to products, and with individuals being able to obtain sizeable amounts of seed or credit).

The WIADP staff wondered if the errors it had observed here and elsewhere of excluding the technical information about growing soyabeans, of giving women only a small amount of seed, and of the extension service operating hypothesis that women only gardened and did not farm, could be remedied. The WIADP was located at the major agricultural research station in the country where soyabean trials were conducted annually, where the **Rhizobium** inoculant for soyabean was produced, and where the technical circulars on soyabean for extension agents had been written. Soyabeans were chosen by the WIADP to remedy the above errors because of the interest already expressed in the crop by the MOA refresher courses. The WIADP posed some questions for study: Were women interested in learning correct husbandry practices? Could they participate in agronomic demonstrations and trials? Were they just helpers or domestic workers or were they farmers in their own right?

To determine if the farmers were interested, the WIADP asked the extension agent who had introduced her class to soyabean recipes to call the class for a meeting where soyabean husbandry practices would be demonstrated. Fifty six (88%) of the women attended. (The rest were attending a funeral and sent apologies.) To ascertain what the women knew about growing beans in general and soyabeans in particular, a questionnaire about spacing, fertilizer and inoculum usage, weeding, pest and disease management, and other production practices was administered.

Also queried was the women's involvement in other staple and cash crop and livestock enterprises, their experience with credit and inputs, and household labor patterns. It was found that all the women farmers grew unimproved maize, groundnuts, pumpkins, sweet potatoes, beans and indigenous vegetables. Half cultivated hybrid maize; some planted cowpeas, groundbeans, sorghum, sugar cane, exotic vegetables and tobacco. The cropping patterns coincided with the patterns delineated by Hansen during his **sondeo**. Some of the women worked alone (if unmarried or if their husbands had migrated for work). Some worked with their husbands either sharing crop operations or being responsible for specific operations on a particular crop. Other women in households where husbands were present did all the work on some crops (such as beans, groundnuts and local vegetables) and on some fields. Eighty-three percent of the women had used fertilizer previously and 25% fertilized local varieties of maize. Fifty-eight percent relied on their husbands or male relatives to obtain fertilizer on credit. In 17% of the cases both wife and husband obtained credit; 8% of the unmarried women obtained credit on their own. Sixteen percent of the female headed households stopped growing hybrid maize because husbands were away and they could not or did not obtain credit for seed and fertilizer. Thirty-three percent of the women relied on their own and on family labor; 58% hired people (usually women) for weeding, and 8% had permanent laborers.

Subsequently, the women were given demonstrations and information by the WIADP and by the male extension staff on correct plant spacing, use of fertilizer and inoculant, proper weeding, signs of readiness for harvest, and storage procedures. Some aspects of the demonstrations had a hands-on approach; for example, bamboo stalks to measure spacing between rows, ridges, seeds and planting depth were prepared and utilized by the women. The technical information presented by the WIADP at the initial meeting and then again two months later at planting was based on recommendations that had been prepared by the agricultural station for a government sponsored extension circular. Planting on ridges was following because it was the government recommendation. The WIADP obtained certified seed from the research station and was able to give each farmer 800 grams of seed and 1.8 kg of the low nitrogen "S" mixture fertilizer (6-18-6) that was enough to plant ten 10m rows. Farmers were shown how to inoculate the seed with the **Rhizobium** bacteria produced at the research station and mixed it in front of them with water and sugar to form a slurry. The farmers went home with seed, fertilizer, measuring sticks and instructions. The **Rhizobium** inoculum only is viable for 2 to 3 days after it is mixed with the seed. It was anticipated that the rains would start that week; but when the rains were delayed, the farmers had to be called back two weeks later to reincoculate their seed. Thirty-nine women farmers returned for the reinoculation; the others had already planted the seed in dry soil.

Six to eight weeks later, a sample of 23 of the 59 female demonstration farmers were visited in their fields. Observations on the growth of the plants were made with the assistance of the extension staff. The farmers were questioned as to when they planted, their husbandry practices, and other aspects of their farming systems. Most planted two weeks after the second inoculation. Field observation revealed that the plants on half of the plots were too small, although most farmers felt no major problems existed. During the growing season the difference in performance between individual farmers was quite noticeable. Farmers had laid out the demonstrations in a variety of ways, and spacing and plant populations varied. In the plots with plants without luxurious top growth (where plants were widely spaced), the canopy did not reach full ground cover and the ridge spacing was too far apart. These farmers experienced pest problems. However, among the plots with good growth, insects were a minor problem, because termite damage occurred after pod formation. The farmers were diligent at weeding which was evident by the low frequency of weeds of the soyabean plots compared with surrounding fields.

The data on farming practices showed that in half of the married households, the wife supplied nearly all the labor on the soyabean plots; husbands helped with the plots on the remainder even though they had not attended the instruction sessions. Unmarried women (only 9% of this sample) did all their own work. When questioned as to proper soyabean husbandry practices, 75% of the farmers in the sample knew the correct spacing and half knew which fertilizers were appropriate. All farmers understood the correct number of weedings and 75% knew which animal manures to use if no commercial fertilizer was available. Less than half of the women grasped the function of the **Rhizobium** inoculum (which is to produce nodulation and increase yield) although two thirds understood how to prepare it.

Results were measured in April 1982 using two plots from each demonstration. Plants were counted and threshed in the field. The cleaned seed was weighed and a 250 gram sample was brought to the research station to adjust the yield for moisture content. Table 7.2 shows the average yield for a sub-sample of 11 farmers and can be grouped into three yield categories of high (2,530 to 2,900 kg/ha), medium (1,160 to 1,400 kg/ha) and low (320 to 660 kg/ha). The reasons for the yield differences are most likely due to variation of plant spacing, inoculation and soil fertility. The farmers who returned to have their seed reinoculated had higher yields than those who did not. Farmers with higher yields had higher plant population densities.

Table 7.2
Yield Characteristics from On-Farm Soyabean Demonstrations in EPA 2 of Lilongwe RDP in 1981-82.

Yield Class	Farmer's Name	Soyabean Yield (kg/ha)	Plant Population (plants/m)	In-Row Plant Spacing (cms)	Inoculated Twice
High	Chembe	2,900	30	7	Yes
High	Unit Centre	2,530	26	9	Yes
Medium	Bau	1,400	32	6	Yes
Medium	Benesi	1,210	28	7	Yes
Medium	Kazola	1,200	14	16	Yes
Medium	Baitoni	1,160	27	8	Yes
Low	Davisoni	660	20	9	No
Low	Chinoko	590	9	24	No
Low	Kabwalo	460	18	11	No
Low	Chauya	400	18	10	No
Low	Kabvala	320	11	18	No

After harvesting the demonstrations, the WIADP interviewed farmers in another rural development project in LADD who had been growing soyabean for some years in order to obtain a greater perspective on farmers' knowledge of the crop and of their cultural practices. Both men and women were growing the crop and their experiences allowed some comparisons. It was found that the men became interested in growing soyabeans as a result of taking agricultural training courses or because of their work experiences with the crop on agricultural estates. The women learned about the crop through home economics courses or from their relatives. The men learned production practices during training courses, however, the women learned only recipes. None of the farmers knew of the recommended **Rhizobium** inoculant or that fertilizers should be low in nitrogen; they said these topics were not discussed in the training courses. The farmers received their seed at a training course (only in some previous years had the seed been inoculated). The men received two to five times as much seed as the women. Most of the women consumed the crop they grew in the first year and did not save any for seed. Some of the men had seed for subsequent plantings. In asking farmers about the labor to grow the crop, 80% of the men said their wives helped and 50% of the women said their husbands helped in growing the crop; however, women threshed and cleaned the seed in all cases. No conscious rotation pattern was known by the farmers, although some grew maize alternately with soyabeans. Thirty-eight percent of the women were interplanting with maize, planting 3 soyabean seeds in between 1 m spaced maize plants. Farmers were interested in the crop mainly as a food for home consumption for porridge, milk, fried snacks and flour to make baked goods. (Malawians do not like the cooked soyabeans as they do not mash well enough to be used as a relish or a sauce.) Two farmers used soyabeans for feeding dairy cows or chickens. Some of the men were attempting to grow the crop for sale, but experienced marketing problems either because no government markets were available in the area or because they did not consider the price at local or distant government markets favorable.

Results from the First Cycle of Soyabean Demonstrations

Based on observations, interviews and discussions with the extension staff and farmers, a number of problems were identified. The first difficulty noted was that many farmers and extension agents alike were having trouble understanding research station recommendations concerning the cultivation of soyabeans. The extension circular was too technical and was based on trials carried out under research station conditions and probably on soils where the **Rhizobium** bacteria was already established. Extension staff noted that the circular did not mention the low nitrogen requirement for inoculated soyabeans. Second, one of the greatest soyabean production problems in Malawi is that the proper strain of **Rhizobium** bacteria is not indigenous to the soil, as it is for groundnuts and cowpeas. The **Rhizobium** is only prepared at one research station using only one method that requires refrigeration. Development projects and farmers needing small amounts of the inoculant had to wait until large estates and the National Seed Company obtained their supply. The LRDP and other projects required inoculant for demonstration plots, training classes and farmer's fields, but they had to cope with the problems of access to the prepared **Rhizobium** as well as their own problems of keeping the **Rhizobium** packets refrigerated (the Units or EPAs do not have refrigeration) and transporting them to the rural areas. A third problem was the small size of the soyabean plants observed in farmers' fields. This was related to lower than optimum soil fertility as well as to the need for inoculant. The WIADP agronomist also postulated that on lower fertility soils, closer ridge spacing might aid in canopy cover. These three problems were independent of gender and affected men and women farmers alike.

The fourth problem was an error in planning and in targeting farmer populations. Women farmers needed to be taught husbandry practices and technical information concerned with production, to be self-sufficient with any new commodity introduced. Female extension staff needed to receive adequate training to teach women how to cultivate a new crop. Male extension staff had to be willing to teach groups of women in their areas either as a supplement to the classes of the female extension staff, or in their own agricultural classes and village meetings. Male staff also could generate interest in new food crops among men taking training courses. Fifth, the amount of seed distributed to class participants, especially to women, was insufficient and uninoculated. Women farmers scarcely had enough to prepare the recipes and few could save seed for the next season. These problems were gender-specific and women were disadvantaged compared with men.

The WIADP took the following actions. With the assistance of several Malawian scientists, a simplified version of the recommendations for growing soyabeans was produced. In addition to an English version, a version was translated into the vernacular (Chichewa) and both versions were distributed to extension workers of both sexes as well as to the demonstration farmers (although many were illiterate) within the project area. In addition, the WIADP in conjunction with some technical officers wrote a syllabus that extension personnel could use for teaching a course on soyabean production to farmers. The syllabus included history of the crop, uses, recipes, botany and general agronomy. Lessons for each topic contained a specific message plus practical activities and the visual aids needed for the class. Finally, female extension agents received instruction in soyabean agronomy from the WIADP and faculty of the agricultural college during the refresher held the following year.

However, improved methods of maintaining viable inoculum were needed because of the problems of distributing viable inoculum and of being able to plant the seed within several days after inoculation. Although this problem has been solved elsewhere using other technologies, the WIADP wondered if it were possible to make do with the existing research station methods, since the method was working in the commercial sector. That is, the **Rhizobium** could be manufactured as before and placed in packets that still required refrigeration. But rather than the extension agent or the farmer mixing the **Rhizobium** with water and coating the seed, one possibility then was to mix the inoculum with moist sand (the Granular method) that was mixed with the seed at the time of planting. This would prolong the **Rhizobium**'s viability until distribution to the farmers. Trials were designed to compare no inoculum, the standard slurry method of coating the seed, and the Granular method of **Rhizobium** inoculation. (In the slurry method, the inoculant is mixed with a 5% sucrose solution to produce a "slurry" and the seed is coated with the mixture. When dried, the seed must be planted into moist soil within three days. The Granular method, which had been tested on the research station for several years but had not yet been recommended, incorporated the slurry into moist sand. The mixture of moist sand and inoculum is viable for a week or so and is placed in the seed furrow at the time of planting.) It was hypothesized that the **Rhizobium** bacteria would persist longer in moist sand than when mixed with the seed. Since most farmers would plant less than the 25 kg. of seed recommended for one packet of inoculum, it was hypothesized that the inoculum should be mixed with sand at the Unit or EPA centers and then distributed to farmers.

Soyabean Trials Testing Methods of Inoculation in 1982-83

At meetings between the LRDP project staff and the WIADP, it was decided to carry out the above treatments with twenty women farmers in four different areas of LRDP. Each of the three treatments was to be planted on six 10m ridges. Two meetings were to be held with the farmers: the first was to serve as an orientation to the trials and to carry out a survey of the farmers' cropping practices; the second was to have the farmers in each area plant a trial at the Unit/EPA center as a group and then to distribute the inputs. Three objectives for the trials in 1982-83 were specified by the WIADP and the LRDP staff. The first was to compare the growth of soyabeans as affected by no inoculum, inoculum mixed with sand, and inoculum mixed with seed. The second was to help popularize soyabeans. The third was to provide an example of how to organize women farmers with extension staff in order to field-test ideas derived from agricultural research stations and/or to solve problems identified by working with farmers.

In December 1982, twenty women farmers participated in planting the trial gardens under the instruction of the WIADP agronomist (Smith 1983) and local male extension staff. A short questionnaire was administered to each farmer regarding her experience with soyabeans, and with other crops and inputs. (The farmers were selected by the extension staff rather than recruited based on interest. None had grown the crop before;

some had no idea how to prepare soyabeans because they did not receive this information in their particular home economics course or they had not taken any course.) To begin the trials, farmers were taught how to form the desired spacing of 18 ridges with 90 cm between ridge centers. Farmers were asked to place the bags of seed on the appropriate groups of ridges. After each farmers was successful in plot layout, she participated in planting seed. Using the bamboo sticks, the ridges were flattened to accommodate two rows of soyabeans. Seeds were planted within the row at a distance of 10 cm (the desired spacing this time), with each woman checking the in-row distance with the 10cm bamboo stick. After planting the treatments with no inoculum and with the inoculated seed, each woman was taught how to apply the sand mixed with inoculum. Every farmer was checked for proper technique in applying seed and sand. Then each farmer received 10 m of string, three bamboo sticks of 90 cm, 25 cm and 10 cm, and the packets of seed in order to plant her own trial the following day. The local extension staff was requested to draw a diagram of each trial so as to be able to locate each of the three treatments. As the trials were to be completely farmer managed, the same order of treatments for each trial was used to avoid confusing the farmers. These were: 1) no inoculum, 2) inoculum with seed, and 3) inoculum with sand.

Results from the Soyabean Trials

About 6 weeks after planting, each soyabean plot was evaluated for nodulation (a sign that the **Rhizobium** inoculation was successful) at the time the plant flowered. At this time it was evident that the first objective of the trials would not be successful. None of the treatments in the trials had good nodulation. The reason for the lack of nodulation is not positively known. Perhaps the **Rhizobium** was inviable prior to inoculation due to overheating resulting from a failure of the refrigeration room where stored after being produced. The warm temperatures after production may have destroyed viable **Rhizobium.** This situation reinforced the WIADP notion that other means of preparing and storing inoculant should be considered.

In spite of the lack of recorded nodulation six of the trials were harvested. The average yield for all plots was 640 kg/ha with a standard deviation of 170 kg/ha adjusted to 10% moisture (See Table 7.3). The low yields achieved in 1982-83 are probably explained by the lack of nodulation in all three treatments and are similar to the yields achieved by demonstration farmers whose seed did not have viable inoculum the previous year. In summary, the experiment was not successful at distinguishing differences in nodulation and yield due to no inoculation, slurry inoculation and granular inoculation, and therefore failed to solve the low yield problem that farmers and extension staff experienced. The lack of nodulation was probably due to inviable **Rhizobium.** The failure of the experiential aspects were probably the fault of the researchers, not of the farmers and extension staff.

Table 7.3
Soyabean Yields from Six On-Farm Trials in EPA 2 of Lilongwe RDP Testing Three Inoculum Methods in 1982-83.

		Yields in Kilograms per hectare*			
Unit	Farmer Name	No Inoculum	Inoculum With Sand	Inoculum With Seed	Farmer Average
6	Chikuse	630	820	530	660
6	Erasimu	790	780	460	680
8	Paliyani	470	380	540	460
8	Lote	800	770	580	720
9	Majinga	360	590	570	510
9	Unit Centre	790	980	730	830

* Adjusted to 10% moisture content.

Average Yield = 640 kg/ha. Standard Deviation = 170 kg/ha.

The second objective of helping to promote soyabeans was successful since the trials of all cooperating farmers did mature a crop, the only soyabeans being grown in LRDP that year. The third objective of demonstrating how to organize women farmers with extension and research staff to field-test new ideas also was successful. Not only did the extension staff assist in organizing women farmers for agricultural research and extension activities, but a more accurate method of instructing farmers in agricultural technologies was devised. At the time of planting the trial at the extension center, each farmer proved able to match the bags for the three treatments. Most farmers were able to repeat the differences between the three inoculation methods. The women were instructed in a laboratory approach, in which each person had to participate in planting the demonstration prior to her own trial. This is different from a demonstration approach, whereby the farmer receives general instructions or where the farmer watches a trial being laid out and then plants her own by herself. (As noted above, the demonstrations the first year showed great diversity when farmers watched and then planted their own). During group instruction, most farmers made mistakes in laying out the bags and planting the seed. Each farmer was corrected by the WIADP agronomist or extension staff until proper technique was acquired. By making mistakes in the presence of a skilled instructor, it was possible to correct errors before the end of the training session.

Further Actions and Conclusions

Many of the questions posed concerning women's involvement in agricultural services have been answered by the demonstrations and trials reported on here. The staff of the LADD and the LRDP learned that women were interested in agricultural subjects if given the opportunity. The staff realized that women could participate in extension demonstrations and research trials. They saw that women were indeed farmers who needed agricultural information in addition to information about cooking and sewing. Interest in agricultural topics could be generated by home economics subjects, but because the number of female extension agents was small and their agricultural training was minimal, there needed to be other ways to provide agronomic information and agricultural inputs to women. The larger and better trained male extension staff were able to work with women farmers and to provide regular extension services to them. Both the development project management and its field workers could target women in their areas as client groups and could adjust their programs to include women. The staff also realized it was possible for extension workers to be retrained in terms of subject matter and methods of dealing with clients. As a result of this new way of thinking, some actions were taken. Thirty percent of the places in LADD's agricultural courses for farmers were reserved for women farmers. The refresher courses for women extension agents began to include information on soyabean agronomy as well as information on the recipes, and in fact, the agricultural content of the refresher courses for women agents was increased. In addition, it became possible for women agents to attend some of the refresher courses held for men, essentially integrating the two groups for the first time. After the trials, the WIADP prepared an extension circular entitled "Reaching Female Farmers through Male

Extension Workers" that was printed by the Ministry of Agriculture (MOA 1983) and that was distributed to all extension staff in the country. The circular legitimated the male staff's work with women farmers in terms of farmer visits, demonstrations, clubs and credit, and offered techniques for working with women farmers. The WIADP helped design new monitoring and evaluation formats for LADD's extension workers and project management that measured extension contacts to both men and women, since previously the forms did not differentiate by sex of farmer. All these changes may be pointed to as part of the effect that the WIADP and the demonstrations and on-farm trials had on the LADD (Spring, Smith and Kayuni 1983a).

Why was the WIADP able to effect these and other changes? Part of the answer lies in the fact that the director of the WIADP was an anthropologist familiar with the local culture. In addition to using a standard FSR approach with a multidisciplinary team, the project benefited from many standard anthropological techniques that were routinely applied. In this exercise and in others that the WIADP conducted, the ways in which "resources are culturally organized and employed" in terms of the division of labor by sex, the allocation of land and inputs, and the food preferences and choices were critical to understanding how to introduce the new technology (Cernea and Guggenheim 1984:7). The use of informal surveys, key informants, and participant observation to learn about farmer's and extension staff's technical knowledge and problems was routine. Once indigenous ideas were considered, difficulties with the new technology could be pinpointed easily. The consideration of structural situations lead to understanding the problems that rural development projects were having understanding gender issues and targeting women as agriculturalists, that extension staff were having with research recommendations, and that farmers were having with extension personnel. Finally, the suggestions for change that were offered took the cultural norms and values into consideration. The rules by which the male staff could work with women farmers and by which extensionists could be reoriented and retrained were appropriate to the cultural situation.

What about the errors that occurred? There were several categories of error; technical, structural and situational. The technical kinds of errors can occur during any Farming Systems Research project. First, the farmers might not truly understand the planting instructions and layout, as was the case in the soyabean demonstrations the first year. Other family members not present during instruction sessions might do some or all the work, as for example some husbands who did not attend the instruction session. In other FSR trials in Malawi husbands were selected as trial farmers, but their wives, who did not attend the training session, did much of the work. The use of surveys and participant observation allowed this error to be known and corrected by the WIADP during the following year. Second, the primary technical error in the second year was with the technology itself. The inoculant was defective because of manufacturing errors and this resulted in the major problem for smallholders not being solved during that planting season. Research now has to consider the problem again.

Another group of errors was structural. First, because of the lack of understanding of the sexual division of labor and women's role in agricultural production, inappropriate models were being used by the extension service and errors were made in targeting the appropriate groups of farmers in development projects. Men and women were given differential training based on stereotypes rather than on productive roles. Women were not targeted as farmers or trial participants but rather for their domestic roles only. Hence, they had reduced access to new technologies and their farming problems were not known. Second, soyabeans were popularized through courses in cooking and nutrition and a minor demand was created. But the production aspects were not set up: little or no seed was available; inoculant and proper fertilizers were difficult or impossible to obtain; and the commercial aspects of smallholder production were not fully addressed. These aspects affected both women and men. Farmers were intrigued with the new crop, but the technical support was lacking. Extension had difficulty understanding research recommendations, and researchers did not know the problems experienced by the farmers and extension workers. Researchers were committed to particular methods of planting and of inoculant preparation and administration that were problematic for smallholders.

The final type of error concerned the mistake made by some Malawians and some expatriate technical assistants of thinking that the WIADP was **only** interested in soyabean production or that the WIADP staff thought that soyabeans were a priority crop for research. In fact, the soyabean demonstrations and trials were only a small part of the WIADP's activities and were chosen because of the MOA refresher courses. The WIADP was attempting to show that some problems were gender specific and some were not; the soyabean demonstrations and trials provided a vehicle for this attempt. The topic was interesting to the WIADP, because of the aspects of home economics and agricultural training. Soyabeans had been selected by the Women's Programs and Food and Nutrition sections of the MOA to improve diet, but the production aspects in terms of smallholders had not been considered. Fortunately, the confusion was resolved when the WIADP prepared better information about its work and disseminated this to people in research and extension.

Finally, it is important to point out successes and changes that occurred as a result of the events described here. First, as noted above, it was shown that women were agriculturalists and interested in new technologies. Second, a precedent for extension and research interacting with each other and with farmers was set up, and technical information was rewritten with the farmer in mind. Third, the method of instructing farmers in planting trials by doing demonstrations first, and being corrected as they went along was noted as alleviating the major sources of farmers' errors. In sum, the purpose of FSR is to keep the people in mind and to correct errors and improve farmers' productivity, income and quality of life. These techniques only work if the errors can be admitted openly and if the appropriate corrections can be made.

Notes

1. The recommended in-row spacing was one plant per 5 cm. The recommendations from the research station stipulated a desired distance between the centers of adjacent ridges as 90 cm, with 25 cm between the two rows of soyabeans on one ridge. It should be mentioned that many agronomists outside Malawi plant soyabean on the flat rather than on ridges and that they use different spacing and inoculum procedures. However, the WIADP was trying to work with existing recommendations. In fact, the recommended spacing was problematic for the farmers here because they have been advised to plant maize, groundnuts and tobacco on ridges ranging from 100 to 120 cm apart. Only the Unit Centre demonstration achieved the desired ridge width of 90 cm.

Bibliography

Cernea, Michael and Scott Guggenheim. 1984.
"Anthropology and Farming Systems Research: A Rejoinder". Paper presented at **The Role of Rural Sociology - Including Anthropology - in**

Farming Systems Research and Technology Generation and Adoption. ARPT/CIMMYT Networkshop, Lusaka, Zambia: November 1984.

Clark, Barbara. 1975.
"The Work Done by Rural Women in Malawi". **Eastern Africa Journal of Rural Development,** 8(2):80-90.

Gladwin, Christina and Kathleen Staudt. 1983a.
"Reaffirming the Agricultural Role of African Women in Household Economics and Rural Development." **Proceedings of the Association of Faculties of Agriculture in Africa: Workshop on the Role of Women in Household Economics and Rural Development in Africa.** Alexandria, Egypt: Association of Faculties of Agriculture in Africa.

——————. 1983b.
"Women's Employment Issues: Discussion." **American Journal of Agricultural Economics,** 65(5):1055-1057.

Gladwin, Christina, Kathleen Staudt and Della McMillan. 1984.
"Reaffirming the Agricultural Role of African Women: One Solution to the Food Crisis". **Proceedings of the Association of Faculties of Agriculture in Africa. Fifth General Conference on Food Security.** April 1984.

Hansen, Art. 1983.
"Farming Systems and Farmer Decision-Making Survey: Tour of Lilongwe Land Development Project." Gainesville, Florida: LRDP. Mimeo.

Hansen, Art and J. Ndengu. 1983.
"Lilongwe Rural Development Project Cropping Patterns: Information from the National Sample of Agriculture." Gainesville, Florida: Farming Systems Analysis Section UF/USAID. Mimeo.

Kinsey, Bill. 1973.
The Lilongwe Land Development Project: A Review of the Background, Accomplishments and Transferability of the Experience of One of the Three IDA-Funded Rural Development Projects in Malawi. Washington, DC: IBRD.

Kydd, Jonathan. 1982.
Measuring Peasant Differentiation for Policy Purposes: A Report on a Cluster Analysis Classification of the Population of the Lilongwe Land Development Programme, Malawi, for 1970 and 1979. Zomba, Malawi: Government Printer.

Lele, Uma. 1975.
The Design of Rural Development: Lessons from Africa. Baltimore: Johns Hopkins University Press.

Mead, Margaret. 1973.
"A Comment on the Role of Women in Agriculture." In I. Tinker and M. Bramsen (eds.). **Women and World Development.** New York: Overseas Development Council.

MOA (Ministry of Agriculture). 1983.
"Reaching Female Farmers Through Male Extension Workers." Government of Malawi: Extension Aids Circular 2/83.

NRDP (National Rural Development Programme). 1977.
National Rural Development Programme. Zomba, Malawi: Government Printer.

NSO (National Statistical Office). 1982.
Preliminary Report: National Sample Survey of Agriculture for Customary Land, 1980/81. Zomba, Malawi: Government Printer.

Smith, Craig. 1983.
"The WIADP Soyabean Programme in the Lilongwe Rural Development Project." WIADP Report 20.

Spring, Anita. 1981.
"Farm Home Assistants and Agricultural Training". WIADP Report 1.

——————. 1983.
Priorities for Women's Programmes. Final Report submitted to the Government of Malawi, Ministry of Agriculture and USAID/Office of Women in Development.

——————. 1984.
Profiles of Men and Women Smallholder Farmers in the Lilongwe Rural Development Project, Malawi. Final Report. Washington, DC: USAID/Office of Women in Development.

——————. 1985.
"The Women in Agricultural Development Projects in Malawi: Making Gender-Free Development Work". In R. Gallen and A. Spring (eds.) **Women Creating Wealth: Transforming Economic Development.** Washington DC: Association for Women in Development.

Spring, Anita, Craig Smith and Frieda Kayuni. 1983a.
"An Evaluation of Women's Programmes in Lilongwe ADD: How LADD Sections and Projects

can Incorporate More Women Farmers in Their Programmes." Gainseville, Florida: WIADP report.

———. 1983b. **Women Farmers in Malawi: Their Contributions to Agriculture and Participation in Development Projects.** Final Report. Washington, DC: USAID/Office of Women in Development.

8

Farming Systems Research in Phalombe, Malawi: The Limited Utility of High Yielding Varieties

Art Hansen

Farming Systems Research (FSR) methodology requires multidisciplinary teamwork and reality testing. Multidisciplinary understanding and teamwork are difficult to achieve for many reasons, including the communication problems and irritations caused by scientists questioning each others' assumptions about research priorities and procedures. Each discipline possesses its own traditions, as do the various international and national agricultural research and extension communities. Teamwork requires and rewards the questioning of each others' assumptions, although these are defended and attacked emotionally as well as rationally.

The author, an anthropologist, initiated and then directed an FSR program in Malawi for more than two years (1981-1983). Experience in Malawi confirmed the importance of learning from farmers, the manifold utility of rapid reconnaissance surveys and on-farm testing, the heterogeneity of local farmers and the need to continue learning about and categorizing farming systems and recommendation domains. Work in Phalombe is reported here, particularly the results of an on-farm trial during the 1981-82 rainy season. "Local" maize and the recommended high yielding variety (HYV) were tested, as was fertilizer. The planned inclusion of low resource and women smallholders in the trial broadened the range of farmers sampled, since earlier on-farm trials had only utilized "good" farmers who could care for trials.

The trials revealed, unexpectedly, the limited utility of the HYV, although HYVs are a standard component of research and extension recommendations in Malawi and elsewhere. Analysis of the Phalombe on-farm trial data showed that the particular HYV was appropriate for "good" farmers in the area, who were a minority. For the majority of farmers the HYV yielded no more and sometimes less than the existing local maize.

Therefore, smallholders were divided into two recommendation domains, one for which the HYV could be recommended and the other for which it could not. All of this sounded rational and scientific, but arguments and emotional displays among research and extension staff revealed the normative quality of the HYV tradition. This experience with maize and farmers provided another lesson about the power of scientific traditions (Shils 1981).

Research in Phalombe

For more than two years (1981-1983) I worked as a member of a multidisciplinary technical assistance team (TAT) implementing a USAID-supported agricultural research project in Malawi. The five year project (1980-1985) was designed to support and upgrade the Department of Agricultural Research (DAR) in the Ministry of Agriculture. A major component of the project was funding advanced professional training of Malawian research staff. The expatriate TAT continued research activities left by Malawians who were away for further training and initiated several new research programs. Other members of the TAT included two agronomists, an animal scientist, a horticulturalist, a plant breeder and an agricultural economist. Each had responsibility for supporting and strengthening research in his own specialty. My responsibility as the farming systems analyst was to initiate a nationwide FSR program and direct a unit within DAR in administering the program (Hansen 1985).

Several years before our arrival, Malawi had administratively decentralized its agricultural program into eight Agricultural Development Divisions (ADDs), each with considerable autonomy. Each ADD was subdivided into project areas, many of which had received separate development funding from various international agencies and had become functioning Integrated Rural Development Projects (RDPs). RDPs were the basic agricultural development units because of their separate funding.

Phalombe RDP, one of the first sites selected for the FSR program, is located in southeastern Malawi between Mount Mulanje to the south and Lake Chilwa to the north; the international border with Mozambique is the eastern boundary. After several months of reviewing background documents and arranging schedules, a rapid diagnostic survey was conducted in May 1981.

Land, rainfall, crops and livestock

Much of the area is a colluvial plain at an altitude of 600 to 700 m, but the flatness of the plain is broken by a number of steep sided rocky hills and mountains, and the southern half is dominated by the towering bulk of Mount Mulanje which rose to 3000 m. This mountain cast most of Phalombe into a rain shadow. The more fertile and well drained pediment soils surround the mountains and hills, while the plain is more variable in drainage and soil texture, ranging from coarse sands to heavy clays. To the north and northwest the plain slopes to Lake Chilwa and the Phalombe River drains into the lake. Heavy clay soils in these lower areas are seasonally or permanently waterlogged (ADD Land Husbandry Unit).

Crops were more important than livestock to Phalombe smallholders, and maize was the most important crop (Table 8.1). More than 75% of cultivated land was devoted to maize, almost all of the local type, and usually grown as the major crop in fields intercropped with one or more of the following: pulses, groundnuts, millets, sorghum, cassava or cucurbits (ADD Evaluation Unit). Cowpeas and pigeonpeas were the major pulses. Staple crops were sometimes densely intercropped, but the most common pattern was for a single staple to be dominant with a number of more sparsely intercropped pulses and cucurbits, and often a sparsely intercropped second staple or a cash crop such as sunflowers, grams and chickpeas. In more waterlogged areas, rice was sometimes intercropped with maize, the rice being transplanted into the furrows of a maize field after the final weeding (Hansen 1982). Tobacco was grown in the more humid areas, and cotton in the drier. These two cash crops were usually monocropped and were an important source of income for farmers with more land. Farmers with less land preferred to devote that almost entirely to intercropped food for home consumption. Food crops also could be sold for cash, since there was always a local market for food (ADD Evaluation Unit).

Almost one fourth of the households (23%) owned no livestock, while a few owned many animals. Although chickens were the most common animal, owned by almost three fourths of the households (71%), more than one fourth (28%) owned only one chicken each. Fewer than one sixth (14%) of households owned cattle, one fifth (20%) goats, and one tenth (11%) pigs (ADD Evaluation Unit).

Malawi has one rainy season lasting from November through April, and unreliable rainfall is a major constraint to agricultural production and stability in Phalombe. Rainfall varies from location to location with higher and more stable rainfall (1000 to 1300 mm annually) east of Mount Mulanje, lower levels (800 to 900 mm) in the central section and along the western flanks of the mountain, and dropping off towards the north and west. Low rainfall in the central, western and northern sections is compounded by erratic distribution; especially prevalent are February dry spells when the maize cobs are forming. An analysis of rainy pentades (five day units) for four rain stations in the central and western sections shows that rainfall distribution was adequate for good maize production only in one of every four years (ADD Land Husbandry Unit).

Table 8.1
Crops and Intercropping in Phalombe:
1968/1969 and 1980/1981

	1968/69 Acres (a)	% of Total	% Inter-cropped	1980/81 Hectares (b)	% of Total	% inter cropped (b)
Maize	25,800	81	96	29,800	76	92
Pulses	23,600	74	100	5,100	13	100
Millet/Sorghum	18,000	56	92	5,400	14	63
Groundnuts	10,100	32	97	2,100	5	76
Cassava	6,000	19	77	3,000	8	30
Rice	--	--	--	1,200	3	--
Total	31,900	100	86	39,100	100	75

Source: The National Sample Survey of Agriculture 1968/69 statistics refer to a smaller area than the Evaluation Unit Working Papers, Blantyre ADD 1982, which refer to the 1980/1981 Phalombe Project dimensions. This is why the 1968/1969 acreage statistics are so much smaller then the later hectarage statistics. The earlier figures are given as a guide to the extent of multiple cropping and the overlap of land in the various crops.

a) The statistics for other crops than maize include many acres that these crops share with maize and/or other crops, so the numbers sum to more than the 31,900 acres and 100%.

b) Intercropping and the extent of crops other than maize in 1980/1981 are underestimated because intercropped fields are recorded in two categories:
- mixed stands, i.e., more than one major crop in a field, and
- scattered plantings in a field with only one major crop.

Scattered planting hectarage is only listed under the major crop and joined with monocropped fields to form a "pure stand" category that is opposed to "mixed stand". Only the maize statistics were disaggregated to show intercropping. Monocropped maize was 8% of all land in maize; mixed stands of maize and other crops were 43%; and maize fields with scattered plantings of other crops were 49%.

Population and Land

Another constraint for most households was a scarcity of land to cultivate. Malawi is one of Africa's most densely populated countries, and the population density of Phalombe, 121 people per km², was more than twice the national average (Table 8.2). Although the average land holding in Phalombe was approximately one hectare (2.5 acres), more than 60% of the households cultivated less than a hectare, and almost a third less than half a hectare (Table 8.3). Land shortage and the drought-prone climate induced individuals and entire households to emigrate in search of land and employment. Many men traveled to the cities or other countries for work, while many others left only during the rainy season to work on tea plantations in the nearby highlands. In these households only the wives and children remained to cultivate the household fields. Entire households also emigrated to live in other areas, which explained the low average annual population growth rate (1.6%), about half the national average.

Table 8.2
Rural Population Growth in Phalombe and Malawi 1966-1977.

	1977 Population	1966-1977 Increase %	Mean Annual Growth Rate %	1977 Population Density (a)	Women per 100 men
Phalombe	168,500	19	1.6	121	113
Malawi	5,547,500	37	2.9	59	107

Source: Malawi Population Census 1977: Final Report.
(a) This refers to people per square kilometer: 121 per square km = 315 per square mile.

The scarcity of adult men was also shown in the high percentage of households headed by women (37%). Within Phalombe the sections with poorest agricultural potential had the highest percentages of women-headed households (ADD Evaluation Unit). Two thirds of these women were unmarried: single, divorced or widowed. Married women were considered household heads only when their husbands returned less than one night a month. Labor was a serious constraint for households headed by women because each household usually contained only one adult worker, while a household headed by a man usually also contained a wife.

Table 8.3
Landholding Size in Phalombe 1968-1969 and 1980-1981.

1968-1969 (a)	1980-1981
	31% less than 0.5 hectares
41% less than 0.8 hectares	62% less than 1.0 hectares
81% less than 1.6 hectares	83% less than 1.5 hectares

Source: National Sample Survey of Agriculture 1968/69 and Evaluation Unit Working Papers, Blantyre ADD 1982. The earlier survey covered only part of the present project.
(a) These were originally expressed in acreage terms and have been translated by the author: less than 2 and 4 acres, respectively.

Diagnosis

Data collected in the informal rapid survey corroborated the background information presented for review by the ADD and RDP staffs, but the FSR survey's conclusions and priorities differed in some ways from existing policies. The FSR program focused on the poor majority, whose highest priority was ensuring an adequate and more secure supply of staple food. Secondary priorities were more secure supplies of other food crops that complemented the staple in the diet, as well as cash income. All households were involved to some degree with the market, so all had needs for cash. Some of the poorest households sold maize immediately after harvest because of a need for money (for taxes, etc.), even though they knew food would run out later. Capital was scarce, and credit was feared because of the consequences of defaulting. Some more fortunate smallholders had more land, more capital and more security, but the survey emphasized the conditions of the majority.

Extension and credit in Phalombe were based on acre (0.4 hectare) production packages, primarily for cash crops such as the recommended varieties of hybrid maize, tobacco, cotton and rice. These integrated packages of seed and several bags of fertilizer followed DAR recommendations but required a lot of capital or credit. Monocropping was recommended for all crops. The FSR program suggested the need to work with units smaller than 0.4 hectare and to break up packages. For the more fortunate smallholders the present packages were suitable, but mini-packages would permit many more smallholders to participate in the credit program and receive seeds and fertilizers. Since capital and labor were scarce and there was a feeling of insecurity with credit, the most appropriate recommendations were those that permitted a step by step progression with each step not requiring a great jump in resource commitment or risk. The survey also pointed out the difficulty for smallholders to accept production packages that required them to use more labor, particularly during the December-January period that was already a labor bottleneck.

Another suggestion was that Phalombe RDP examine its extension coverage to see whether women smallholders were receiving enough extension advice. Women provided most of the labor on smallholder crops in Malawi (Clark 1975; Spring **et.al.** 1983), but were even more important in Phalombe since more than a third of all households were headed by women. Finally, the survey suggested that the project conduct a vaccination program because chickens were an important source of protein and cash income to many smallholders and were unprotected against Newcastle disease.

In terms of agricultural research, the FSR program focused on improving both the yields and the stability of staple food production within a context of multiple cropping. The primary staple to test was the preferred and most common local staple, maize, followed by later trials with sorghum and cassava. Basic research on maize varieties, fertilizer and husbandry practices did not appear to be needed immediately in Phalombe. Based on years of research, the national maize research program had already recommended composite and hybrid maize varieties and fertilizer packages for semi-arid areas such as Phalombe. What remained to do was test the recommendations under realistic smallholder conditions, i.e., intercropped and managed by smallholders who were representative of the majority.

On-farm trials had been conducted by DAR for years, but they had been used as extensions of the research stations. Malawi encompasses a wide variety of ecological niches. Since the research stations were representative of only a few environments, DAR researchers had used farmers' fields to test ecological adaptation of varieties and practices before recommending them. These trials had always been managed by research staff with assistance from extension. Farmers had provided the land and, often, labor for weeding under research or extension staff guidance. To ensure the success of the trials, the farmers had always been carefully selected by extension or research to ensure that they were "good" farmers who would take care of the trials. Good farmers were those who had demonstrated their progressiveness by adopting agricultural recommendations and who had a lot of experience working with extension.

The on-farm trials suggested by the FSR program were different. Farmers were selected who represented the agricultural and socioeconomic features of the clients, i.e., the poor majority, and the farmers managed the trial. The purpose was not to test ecological suitability but rather to learn how the treatments performed on local farms under smallholder conditions and management and how smallholders reacted to the various treatments.

Maize Traditions

Two maizes were tested, the "local" type and the composite recommended for semi-arid areas named Chitedze Composite A (CCA). Local maize was by far the most common maize grown in Phalombe and throughout Malawi. The National Sample Survey of Agriculture in 1980-81 found that 52% of cultivated land in Malawi was in local maize, while only 6% was in HYVs (NSO 1982). Despite its popularity, research staff had informed me that local types were unimproved and unresponsive to fertilizer. They had been demonstrated to be low-yielding, as shown by years of on-station and on-farm trials. Local maize types were part of the traditional agriculture that research and extension were trying to change.

It took me months of interviewing smallholders around Malawi to confirm that local maize really was a folk category with only two defining characteristics. One was that it was "flint" (hard starch) maize, which was preferred by farmers for home processing (pounding), storage and consumption. The other was that the seed was produced from the farmers' previous crop or was purchased in local markets from other farmers, i.e., the seed did not come directly from the government.

DAR was working with and recommending hybrid and composite HYVs. Although hybrid maizes were higher yielding than composites or locals, the "dent" (soft starch) quality of hybrids led smallholders throughout Malawi to reject them as a food crop for home use. This became obvious during months of interviews as farmers indicated how hybrids and local fitted into their farming systems. Hybrids were usually only grown by smallholders in Malawi as a cash crop for immediate sale after harvest. Many smallholders in more commercialized areas such as Lilongwe RDP grew two types of maize, local maize for food and sale and hybrid exclusively for sale (Hansen and Ndengu 1983).

Improved composites were flinty enough to be utilized for home processing and consumption as were the local maizes. Both research and extension preferred the hybrid with its higher yield, but the DAR maize program had continued research in both hybrids and composites because of the known farmer preference for flinty maize for home use. Since the research priority in Phalombe was improved food crop production, the recommended composite rather than recommended hybrid was chosen for testing against local maize.

Although villagers and research and extension staff implied as much, the English term "local" and the Chichewa term **chimanga cha makolo** (maize of the ancestors) did not actually mean that the maize types originated locally nor exclude the possibility of genetic mixtures through open pollination with maizes introduced by the government. There seemed to be a "localization" process that occurred as governmental, i.e., non-local, origins were forgotten. For instance, in the Lilongwe RDP near the capital where seed exchange and introduction programs had been active for many years, the local maize category appeared to include synthetics and composites that the RDP itself had introduced. As reported by extension staff, smallholders recognized many named types of local maize, and some of the names were those of varieties introduced earlier. Phalombe had been more isolated, development activity more recent, and no equivalent mixing of maize genetic material was obvious.

My research interest in local maize began during the course of FSR diagnostic surveys in Lilongwe RDP before the Phalombe survey. Many smallholders in Lilongwe were fertilizing local maize, and at least a small number were consciously experimenting with fertilizer levels and timing. Sometimes the fertilizer was purchased, but often the smallholders were applying fertilizer to local maize when the fertilizer had been received on credit to be applied to HYV hybrid maize.

There was a major disagreement between the research and extension staffs and myself concerning the meaning of this smallholder behavior. Some members of the FSR survey team interpreted what the farmers were doing as cheating or stealing; smallholders were misappropriating fertilizer resources that should have been applied to the recommended variety. My interpretation was quite different. The farmers were showing they wanted to grow local, that they thought local responded to fertilizer, that applying some of the fertilizer to local maize was considered a better use of the resource than applying all of it to the HYV, and that they were unsure of how much and when to apply fertilizer to local maizes. The FSR diagnostic survey provided all of us with the opportunity to learn from farmers, but scientific assumptions interfered with our ability to really listen. Because of the anthropological tradition I was more prepared to hear what the farmers were saying.

Extension staff admitted that they did not give smallholders any advice about growing local maize. Extension was completely oriented toward the recommended HYVs and working with farmers who had received credit packages. Research staff pointed out that there was an old but still existing recommendation about fertilizer levels for local maize. When I asked for the data on which the recommendation had been based, it could not be found. Consequently, the FSR program in Lilongwe decided to collect a set of local maize samples and test their responsiveness to fertilizer. Trials would be run on farmers' fields, each farmer using his own local maize, with a control trial at the research station.

The decision to study local maize with fertilizer was controversial. Many DAR professionals were surprised, some pleased, others unhappy, a few outraged. Many conversations ensued, and I learned a lot about the history of maize research in Malawi and how breeders in Malawi and elsewhere had spent years improving maize to develop HYVs. Including local types in trials was a waste of time, I was told. It was counter-development, a step backward into the past, and an insult to the breeders and agronomists.

Extension people, on the other hand, were undecided on the issue. In Lilongwe RDP where this issue had first arisen, many extension staff were pleased with my decision because smallholders had been persistent about the advantages of certain named local types. Various extension agents questioned whether it was against ministry policy to do anything with local maize. In Phalombe when I suggested to extension staff that head-to-head confrontations of local and recommended varieties in their demonstrations would help farmers recognize the differences, several extension agents expressed the concern that the recommended varieties might not win such a confrontation, and that would seriously damage extension.

The use of local maize in Phalombe on-farm trials was much less controversial than in the Lilongwe trials, but the background of assumptions, concepts and arguments was important in assessing the impact of the Phalombe trial results. Local maize was used in the on-farm trials in Phalombe because FSR methodology specified including common farmer practices as check plots, permitting farmers and researchers to compare suggested innovations with existing methods (Hildebrand and Poey 1985).

Trial Design

DAR and Bunda College (University of Malawi) agronomists, ADD and RDP staff and Phalombe farmers collaborated in planning the trial, although I assumed leadership and ultimate responsibility. The technical and social scientists first worked out the basic design, then meetings were called in the two Phalombe villages where the trial was conducted, and farmers were consulted. In fact, the trial design was modified to incorporate smallholder suggestions about intercropping density and timing (Hansen **et.al** 1982).

Because the on-farm trial was run under smallholder management, standard agronomic research designs were not used, such as randomized blocks with multiple replications. Smallholders needed to be able to handle the trials and understand the alternatives, so a simple design was used in which each smallholder had only one instance of each treatment. Smallholders were then treated as blocks for the purpose of statistical analysis.

Final plans were a simple two by two factorial test of two maize varieties under two levels of fertilizer, all treatments with the same intercropping mix of maize, cowpeas and sunflowers. One of the four treatments was the recommended composite HYV and fertilizer package. Another treatment was the unimproved and most common pattern in Phalombe, the local maize type without fertilizer. The other two treatments were mixtures of unimproved and recommended; local maize with fertilizer as one treatment, recommended variety without fertilizer the other. These treatments allowed us to test whether maize variety or fertilizer alone or in combination improved yield and stability. We anticipated improvements over local practice but needed to test this under farm conditions.

Fertilizer was applied at the level and in the manner recommended for composite HYV - one 50 kg bag of 20:20:0 and two bags of sulfate of ammonia (S/A) per acre, or 2.5 and 5 bags per hectare. Each of the two fertilizers was 20 to 21% nitrogen (N), giving the recommended level for composites of 30 kg of N per acre. 20:20:0 was administered as a basal dressing at time of planting, and S/A was later added as a side dressing.

The intercropping pattern was fairly common in Phalombe. Most smallholders mixed maize and small amounts of cowpea seeds before planting and then planted them in the same stations. Sunflower was the most commonly grown cash crop in Phalombe and was frequently intercropped with maize. The cowpeas in the trial provided the complementary food (relish crop), the sunflowers a source of cash, and the plot therefore a complete mix of staple food, complementary relish crop and cash. Maize and cowpeas were planted at the same time, three maize seeds per station three feet (0.9 m) apart on ridges three feet apart, while cowpeas were very dispersed following villager recommendations. Sunflower was delayed in planting until after the maize was well established. This reflected the smallholder desire to promote the maize crop and treat other crops as bonuses that were not to endanger the staple. Each of the four treatments was eight ridges wide and ten maize stations long. This was much larger than usual research plots, and large enough for the smallholders to be able to appreciate any differences in yield, labor costs, etc.

Farmer Selection and Education

ADD management gave highest priority to the central section of Phalombe, an area of medium rainfall where half of the population lived. Two villages there were selected, five miles apart with similar rainfall and soil conditions. Public meetings were called in both villages by the RDP extension staff, and the FSR program and trial plans were discussed. Villagers made several suggestions that were incorporated into the trial design, and each village then was asked to supply eight volunteers to host the trials on their own fields. The total sample thus consisted of 16 farmers in two separate villages. Smallholders with little land to cultivate were preferred because they were the ones who needed to intercrop, and the FSR program asked that approximately half be women. As it turned out, the trial farmers were split almost equally between smaller and larger, women and men, and the women between married and unmarried.

In two subsequent public meetings and in visits to their fields, the trial farmers and their neighbors learned more about the design of the trial. Verbal explanations were supplemented with tracing diagrams in the dirt, and many questions were answered before people were satisfied that they understood what was going to happen and why. One part of each farmer's field was subdivided into four plots, each plot being eight ridges wide (7.2 m) and ten maize stations long (9 m). As an additional guide, signs were given to each farmer to mark each plot, and they were in abbreviated Chichewa instead of the usual scientific coding. MAK stood for **chimanga cha makolo** or local maize, CCA for the composite, and -F for the plots with **feteliza** or fertilizer.

Seeds, fertilizers and a regulation cup for applying the fertilizer were distributed, and the farmers were given financial information about the inputs, i.e., prices at the government market and the size of the bags in which they were sold. Correct timing for applying fertilizers to maize was discussed and, when someone commented on the unusual color of the CCA seeds, the reason why they had been coated with a chemical was explained. This turned out to be important later on because, when planting was delayed in one village, one elderly women ate the local maize seed because she was hungry. She then told us she would have eaten the CCA seed also if she had not learned about the poisonous coating.

Monitoring

The farmers provided the actual labor and management for the trial. Each week each of them was visited by a local research or extension agent, and each month the FSR team visited from Lilongwe. Weekly visits noted the dates of significant operations (plantings, weedings, fertilizer applications), natural occurrences (rain, army worms, etc.) and the condition of the treatments. The agents also offered advice concerning the trial plots.

During the monthly visits we inspected the trial plots, interviewed the trial farmers about their reactions and continued investigations into Phalombe farming systems. During each visit the farming systems staff met with RDP management and staff to talk about the progress of the trials. This continual interaction between research and extension was necessary to ensure that extension staff understood the trials and could confidently extend the outcome to local smallholders. In February a special multiple cropping survey was conducted to supplement data collected in the original farming systems survey. Although the 1981-82 trial used a standard intercropping pattern in all four treatments, intercropping was a variable that needed to be investigated in future trials.

Farming was always a risky business, and one of the two villages was especially unfortunate that year. The farmers in that village and we suffered through an early drought that caused the eight farmers to have to replant their maize in late December, a month later than the usual planting time and a damaging blow in an area where the rains often stop early. When the rains came, so did erosion, and two of the trial plots were washed away in that same village because people were cultivating hillside slopes. Army worms that year attacked many areas of Phalombe, including that same village, and there were the usual stalkborers and termites to combat in both villages. We had originally intended not to use any pesticides and to expose the trial plots to the same uncontrolled environment of pests that regular fields confronted. Army worms and extension agents changed our minds when the plots were really attacked, and insecticides were distributed for the farmers to apply. This was one reason why our final yields were higher than local averages.

In May maize was harvested. In each village the FSR staff and trial farmers worked together to harvest all of the plots, so all of the farmers would have the opportunity to see how the treatments responded in various ecological niches. Since smallholders were accustomed to evaluating or measuring volumes rather than weights, yields were measured both ways, using a scale for weight and a standard calibrated tin for volume. After all of the plots were harvested, another meeting was called in each village to discuss everyone's evaluations of the different treatments. Smallholder perceptions have been discussed elsewhere (Hansen **et. al.** 1982). This article concentrates on scientific evaluations and traditions.

Trial Results

Table 8.4 presents the yields from the plots of eight farmers in one village and six in another, two plots being omitted due to erosion and poor germination. The statistics represent metric tons per hectare (MT/ha) of usable grain. Usable grain was defined by the smallholders themselves as they worked with us to shell and weigh the harvest. They eliminated all rotten grain and that which was very badly eaten away by weevils. As noted by DAR, smallholder criteria for defining usable and unusable grain differed from the criteria used by laboratory technicians, who discarded all grain that had been attacked at all by insects (DAR Crop Storage Research Section).

Table 8.4 shows obvious differences: 1) between the means

for villages, 2) among the means for farmers in each village and 3) between the means for the two fertilizer levels in the first village. The differences between the two villages is easily explained. Both villages planted in late November, but the second village did not receive enough rain at that time to sustain the seedlings and had to replant in late December, after their first real planting rains. The first village received enough rain in late November and early December, so their maize had a month head start. The second village also suffered severe attacks by army worms in January, but this was less of a factor than the rain and the time of planting. Yields in the first village, therefore, showed how the treatments responded to better conditions. Yields in the second village reflected the adverse conditions that continually threatened Phalombe agriculture.

Table 8.4
Maize Yields from Phalombe On-farm Farmer managed Trials 1981-1982.
Usable Grain in Metric Tons per Hectare (MT/ha).

	1	2	3	4	5	6	7	8	Mean
				First Village					
Four treatments				8 farmers					
Local Maize (LM)	2.2	2.2	1.9	1.2	1.3	0.9	1.0	0.5	1.4
Fert. Local (LM-F)	3.6	3.7	4.3	3.2	2.3	2.3	3.1	2.8	3.2
CCA Maize (CCA)	3.5	2.0	2.9	0.4	0.6	0.5	0.6	0.3	1.3
Fert. CCA (CCA-F)	5.0	4.7	4.3	3.5	2.4	1.7	3.0	2.8	3.4
Mean for Farmer	3.6	3.2	3.3	2.1	1.7	1.3	1.9	1.6	2.3
				Second Village					
Four Treatments				6 farmers					
Local Maize (LM)	1.8	1.1	1.6	1.0	1.6	0.6			1.3
Fert. Local (LM-F)	3.2	2.5	2.9	1.2	1.9	0.8			2.1
CCA Maize (CCA)	2.2	0.7	0.9	0.3	1.1	0.3			0.9
Fert. CCA (CCA-F)	2.9	2.5	2.1	1.1	0.8	0.4			1.6
Mean for Farmer	2.5	1.7	1.9	0.9	1.4	0.5			1.5

Both maizes responded strongly to fertilizer. Both local and CCA more than doubled their yields in the first village under better conditions, although the effect of fertilizer was not as marked in the second village with its generally poorer performance. To our surprise there was little difference between the two maizes at either level of fertilizer in the first village under better conditions, and local performed somewhat better in the second village under poorer conditions.

Table 8.5
Analysis of Variance - Results for On-Farm Maize Trials.

Source of Variance	Degrees of Freedom	Mean Square	F Ratio**	Significance
		First Village		
Farmers (8)	7	3.947	2.41	94%
Fertilizer (2)	1	25.740	15.74	GT 99%*
Maize Type (2)	1	0.428	0.26	Insignificant
Fert x Maize	1	Insignificant		
Error	21	1.635		
		Second Village		
Farmers (6)	5	2.049	3.53	97%
Fertilizer (2)	1	3.450	5.94	97%
Maize Type (2)	1	1.000	1.72	Insignificant
Fert x Maize	1	0.010	Insignificant	
Error	15	0.581		
		Combined Villages		
Villages (2)	1	11.550	10.83	GT 99%*
Fertilizer (2)	1	25.515	23.93	GT 99%*
Maize Type (2)	1	0.026	Insignificant	
Village x Fert	1	3.676	3.45	93%
Village x Maize	1	1.403	1.32	Insignificant
Fert x Maize	1	Insignificant		
Error	49	1.066		

* GT = "Greater Than"

** In the analyses for the individual villages, the smallholders are used as blocks, and there is only one replication of each treatment per block. In the analysis of the combined villages, the villages are used as blocks, and there are eight replications (farmers) in the first village/block and six in the second.

It was important that the technical scientists accepted the trial results, and the TAT maize breeder was asked for advice about how to proceed with further statistical analysis. He suggested the analysis of variance (ANOVA). This analysis in Table 8.5 apportioned total variance among the treatments, the blocks (farmers), and the individual values (random error) - the higher the mean square the greater the variance attributed to that factor. In analysis of the combined villages, the villages were used as blocks and the farmers as replications of each block. The F ratio expressed the difference between the mean squares of that factor and of error - the higher the ratio the more unlike were the populations. The significance statistic expressed the probability that such a difference was not caused by chance.

The effect of fertilizer was highly significant in the first and in the combined villages, and significant in the second village. Maize types were never significant. The differences among farmers were significant in each village, and the difference between the two villages (explained above by reference to rainfall and time of planting) was highly significant. Differences among farmers will be addressed after first discussing the maize type versus fertilizer issue.

Table 8.6
Smallholder Maize Yields by Fertilizer Application Rate, Phalombe 1980 - 1981.

Fertilizer Application (kgs/ha)	Local Maize Yield (MT/ha)	Number of Plots	UCA Maize Yields (MT/ha)
None	0.8	431	1.8 for 12 plots
1- 49	1.6	5	throughout the
50- 99	1.3	21	Phalombe project.
100-149	1.3	22	
150-199	1.4	19	1.5 when plots
200-249	1.9	6	in one area
250-299	2.0	4	with highest
300-349	1.9	6	rainfall are
350+	1.2	6	excluded.

Source: Evaluation Unit, Blantyre ADD.

The importance of fertilizer to maize yields was obvious, but the insignificant relationship between maize type and yield needed to be examined more closely. Maize type and fertilizer relationships were also studied in yield data collected by the ADD Evaluation Unit for the 1980-81 National Sample Survey of Agriculture. More than 500 smallholder maize plots in Phalombe were sampled during that cropping season, although very few (only 12) plots in their sample grew anything other than local maize. All of those 12 were UCA, another composite similar to

CCA. That data showed a significant difference in mean yield between local and UCA, but that difference appeared to be largely a matter of differential fertilizer application rates. Table 8.6 permits comparisons between the maize types at similar levels of fertilizer. Recommended levels of fertilizer for composites were 7.5 bags (375 kg) per hectare. Local maize yields equalled the mean for UCA once 200 kgs per hectare were applied.

Another important dimension was yield stability. Farmers wanted higher yields and a more secure or stable production. More stability may be defined as reduced variability, and relative stability of the two maize types was shown to some extent in Table 8.4 by looking at the range of yields. The coefficients of variation (c.v.) in Table 8.7 measured the extent to which the individual yield values deviated from the mean. This statistic was corrected for the magnitude of the different means (c.v.= standard deviation divided by the mean), so all of the values in Table 8.7 were directly comparable - the higher the c.v. the more variable and unstable. Except for local in the second village, both maizes were more stable when fertilized, a feature most noticeable in the first village under better growing conditions. The inverse of stability may be defined as responsiveness, and CCA seemed to be more responsive to its environment, whether adverse or favorable.

Both yield and stability improved with fertilizer. Unfortunately, fertilizer was a costly input, and lack of capital and fear of credit inhibited people from high cost inputs. The government marketing agency (ADMARC) sold inputs to smallholders and purchased their maize. Recommended levels of fertilizer cost 61.25 Malawian Kwacha (MK61.25) per hectare in 1981, based on an ADMARC price of MK8.50 per 50 kg bag of 20:20:0 (2.5 bags recommended) and MK8 per bag of S/A (5 bags). REcommended seeding rates for CCA cost MK6.25 per hectare, based on a price of MK2.50 per 10 kg bag (2.5 bags). In 1982 ADMARC bought maize for MK0.11 (11 tambala) per kg or MK110 per metric ton (MT). Thus, a yield increase of 0.6 MT/ha was more than enough to offset the cost of fertilizer, and 0.1 MT/ha would pay the CCA seed costs. Table 8.8 shows that it was profitable to apply fertilizer to both types of maize in both villages, although there was very little average profit in the second village.

Thus far, the analysis had only examined aggregates and means. Differences among farmers were very important and, in this instance, statistically significant (Table 8.5). Table 8.9 illustrates this by plotting the mean yields of unfertilized and then fertilized maize, combining both varieties, for all fourteen farmers. There was no normal distribution tailing off to either or both extremes. Instead, there were essentially two categories of farmers in each of the two villages, a low yielding and a high. This division held true with or without fertilizer.

Table 8.7
Yield Stability as Measured by Coefficients of Variation.

First Village		Second Village		Both Villages	
Treatment	c.v.	Treatment	c.v	Treatment	c.v.
Local Maize	45	Local Maize	35	Local Maize	39
Fert. Local	22	Fert. Local	45	Fert. Local	35
CCA Maize	98	CCA Maize	78	CCA Maize	88
Fert. CCA	34	Fert. CCA	63	Fert. CCA	52

Table 8.8
Profitability of Fertilizer Application to Maize (MT/ha).

Maize Type	First Village		Second Village	
	Yield Increase With Fertilizer	Profit	Yield Increase With Fertilizer	Profit
Local	1.8	+1.2	0.8	+0.2
CCA	2.1	+1.5	0.7	+0.1

Note: Each metric ton is worth 110 Malawi Kwacha at 1982 prices.

Three farmers in the first village (1,2,3) and one in the second (A) distinguished themselves by high yields without fertilizer, producing more than double the yield of the majority of the farmers in the sample. Fertilizer improved all of the yields in the first village, but the same three remained far in front. The second village suffered more from poor rainfall, pests, etc., which probably explained why the first farmer (A) did not maintain his lead.

More ominous was the failure of half of the farmers in the second village (D,E,F) to reap any significant advantage from fertilizer. All three would have lost money by buying and applying fertilizer. The relationship between crop failures or very poor yields and unprofitable returns from applying fertilizer was clear. Even fertilizer was not a safe recommendation for everyone. This probably explained why FSR surveys throughout Malawi had found that smallholders consistently delayed applying fertilizer past the recommended time. They were probably waiting until they were able to judge the rains and the health of the crop before committing such an expensive input.

Table 8.9
Smallholder Yields by Village and Fertilizer Treatment
Maize Yields Represent Averages of Both Maize Varieties (MT/ha).

	No Fertilizer		With Fertilizer	
Range of yields	First Village	Second* Village	First Village	Second Village
4.0 - 4.5			1,2,3	
3.5 - 3.9				
3.0 - 3.4			4,7	A
2.5 - 2.9	1		8	B,C
2.0 - 2.4	2,3	A	5,6	
1.5 - 1.9				
1.0 - 1.4		C,E		D,E
0.5 - 0.9	4,5,6,7	B,D		F
0.0 - 0.4	8	F		

* Farmers in the second village are coded by letters in this table rather than by numers 1-6 for ease of presentation.

Recommendation Domains

Looking at the differences among farmers led to seeing the differential utility of the composite HYV and the existence of two recommendation domains. Although aggregate data showed no yield difference between maize types, adopting the HYV made real sense for the three best farmers in the first village (1,2,3). These three got high yields from both varieties, with and without fertilizer, but they consistently got better yields from the composite and demonstrated the advantage of CCA over local in high-yielding situations. There was an obvious relationship between the "good" farmers and the suitability of the HYV.

While the three good farmers had success with the HYV, the other five farmers in the same village harvested 0.6 MT/ha or less on their plots of unfertilized CCA, severely depressing the mean for that treatment. Their plots showed no differences between maizes or, with the lowest yields, the advantage of local stability. When the added cost to buy the CCA seed was considered, equal to the income from 0.1 MT/ha, it was obvious that the HYV was not recommended for these farmers. Shifting to the HYV would not be profitable for them given their yields. The same conclusion was applicable to everyone in the second village with the possible exception of the first farmer.

The analysis suggested that the majority of farmers in this section of Phalombe would not profit in yield or in stability of yield by adopting the recommended HYV, but the minority of good farmers would. The minority who got high yields from all treatments constituted one recommendation domain, and extension could continue with confidence to advise them to adopt the HYV and fertilizer package. On the other hand, the majority of farmers were a separate recommendation domain. For whatever reason, the "recommended" HYV was not recommendable to this domain at this time. This conclusion and these data have been elaborated graphically elsewhere (Hildebrand 1984; Hildebrand and Poey 1985: 126-132.)

Some factor or factors interfered with yields in general on the fields of the majority of farmers. The yield data (Table 8.4) could not satisfactorily explain the gap in yields between the two farmer categories nor the low yields of the majority because the trial had been designed to measure difference among treatments rather than among farmers. Each farmer was statistically a separate block, and each block had its unique location, microclimate, soils, field history and farmer. For convenience the interference factors were lumped in a "farmer husbandry" category.

Interview data gave some clues. Differential residual fertility was a factor since two of the best farmers (1,2) had planted fertilized tobacco in the on-farm trial fields the previous year. Husbandry was also a factor; some farmers invested more time and care on their fields, while others spent less time, were sick or old and weak, had domestic problems diminished labor availability, etc. The highest yielding farmers (1,2,3,A) were male, not usually high status, usually tobacco growers and had larger amounts of land than the others. These differences among smallholders needed to be investigated in future trials, so that research could isolate and work with the husbandry factors that inhibited farmers from achieving higher yields and advise extension agents on how to recognize farmers who would benefit from different recommendations.

Conclusions

Although the trial had not been designed to identify or discriminate among farmer husbandry factors, the trial had succeeded in testing under realistic smallholder conditions a set of traditional inputs. Local maize grown without fertilizer was a popular tradition, while CCA and fertilizers were scientific traditions (Shils 1981). Reality testing had confirmed one scientific tradition, the use of fertilizers, and provided surprising insight into the differential utility of the two maize traditions.

On-farm testing had shown its scientific usefulness in this instance, but a key had been the selection criteria for the sample of farmers. Most of the Phalombe trial farmers conformed in scale and sex to the majority for whom recommendations were being sought. Previous on-farm trials in Malawi had consistently shown the advantage of HYVs over local maizes, but the farmers in those trials had always been good farmers. This trial included a broad range of smallholders and showed that the best farmers in Phalombe were a separate recommendation domain. The previous smallholder samples in other on-farm trials had not been representative of the majority, but DAR had assumed that results from those trials were applicable to all smallholders. One implication of the Phalombe results was that agricultural researchers had to question that assumption. HYVs had been assumed on the basis of previous trials to be recommendable to all farmers. That assumption now had to be questioned as well.

Previous variety and husbandry recommendations in Malawi had been very general, many at a national level and others subdivided by ecological factors such as rainfall, altitude and temperature. Research and extension had known that the recommendations had to be drawn more exactly, and DAR was in the long process of doing that in coordination with the ADDs and RDPs. Only biological and climatic factors were being considered, however, not socioeconomic differences among farmers. The Phalombe trial demonstrated the importance of intra-locality variation among smallholders and the usefulness of the FSR approach. Socioeconomic variation needed to be considered in revising agricultural recommendations.

FSR concepts and methods are compatible in many ways with anthropological traditions, and anthropologists and our accumulated body of literature have much to offer (Hansen 1984). This case demonstrates a successful test of various technical assumptions and a contribution to the knowledge of Malawi smallholders using the FSR approach. The FSR program was directed by an anthropologist, but this research calls for many skills and insights, and only multidisciplinary teamwork and participation by farmers can provide that. My professional colleagues taught me a lot during this process, and we all learned from the farmers. As do all scientists, anthropologists have assumptions and traditions, and they are severely tested by other scientists in FSR teams.

Bibliography

ADD Evaluation Unit. 1981, 1982.
Working Papers. Blantyre, Malawi: Blantyre Agricultural Development Division. Mimeo.

ADD Land Husbandry Unit. n.d.
Geography and Geomorphology. Blantyre, Malawi: Blantyre Agricultural Development Division. Mimeo.

Clark, Barbara. 1975.
"The Work Done by Rural Women in Malawi." **Eastern African Journal of Rural Development** 8(2); 80-91.

Collinson, Michael P. 1982.
"Farming Systems Research in Eastern Africa: The Experience of CIMMYT and Some National Research Services, 1975-81." International Development Paper #3. East Lansing, Michigan: Department of Agricultural Economics, Michigan State University.

DAR Crop Storage Research Section. 1980.
Annual Report 1979/80. Chitedze, Malawi: Department of Agricultural Research. Mimeo.

Fresco, Louise. 1984.
"Comparing Anglophone and Francophone Approaches to Farming Systems Research and Extension." Networking Paper #1. Gainesville, Florida: Farming Systems Support Project, International Programs, University of Florida.

Gilbert, Elon H., David W. Norman and F.E. Winch. 1980.
Farming Systems Research: A Critical Appraisal. Development Paper #6. East Lansing, Michigan:

Department of Agricultural Economics, Michigan State University.

Hansen, Art. 1982.
"Intercropping and Farming Systems in Three Areas of Malawi." In **Proceedings and Materials from the Conference on Intercropping Research in Malawi, 20 October 1981, Chitedze Agricultural Research Station.** A. Hansen (ed). Chitedze, Malawi: Malawi Agricultural Research Project. Mimeo.

———. 1984. "Zambia, Farming Systems Research and the Anthropological Body of Knowledge." Keynote address to the Networkshop on the Role of Rural Sociology (including Anthropology) in Farming Systems Research, held 27-29 November 1984. Lusaka, Zambia. Mimeo.

———. 1985. "Learning From Experience: Implementing Farming Systems Research in the Malawi Agricultural Research Project." In **Domestic Farming Systems Conference; Proceedings.** E.C. French and R.K. Waugh (eds). Gainesville, Florida: University of Florida Press.

Hansen, A., E.N. Mwango and B.S.C. Phiri. 1982.
Farming Systems Research in Phalombe Project, Malawi: Another Approach to Smallholder Research and Development. Gainesville, Florida: Center for Tropical Agriculture, University of Florida, in cooperation with Farming Systems Research Section, Department of Agricultural Research.

Hansen, A. and J.D. Ndengu. 1983.
"Lilongwe Rural Development Project Cropping Patterns and Information from the National Sample Survey of Agriculture." Paper presented to evaluation officers at the Ministry of Agriculture, April 1983. Lilongwe, Malawi. Mimeo.

Hildebrand, Peter E. 1984.
"Modified Stability Analysis of Farmer Managed, On-Farm Trials." **Agronomy Journal** 76;271-274.

Hildebrand, P.E. and F. Poey. 1984.
On-Farm Agronomic Trials in Farming Systems Research and Extension. Boulder, Colorado: Lynne Rienner Press.

Ministry of Agriculture. 1981.
Guide to Agricultural Production in Malawi 1981/82. Lilongwe, Malawi: Extension Aids

1981/82. Lilongwe, Malawi: Extension Aids Branch, Ministry of Agriculture.

National Statistical Office. 1970. **National Sample Survey of Agriculture 1968/69.** Zomba, Malawi: Government Printer.

———. 1980. **Malawi Population Census 1977: Final Report.** Zomba, Malawi: Government Printer.

———. 1982. **Preliminary Report: National Sample Survey of Agriculture for Customary Land 1980/81.** Zomba, Malawi: Government Printer.

Norman, D.W., E.B. Simmons and H.M. Hays. 1982. **Farming Systems in the Nigerian Savannah: Research and Strategies for Development.** Boulder, Colorado: Westview Press.

Schultz, T. W. 1964. **Transforming Traditional Agriculture.** New Haven, Conn: Yale University Press.

Shaner, W.W., P.F. Philipp and W.R. Schmehl. 1982. **Farming Systems Research and Development; Guidelines for Developing Countries.** Boulder, Colorado: Westview Press.

Shils, Edward. 1981. **Tradition.** Chicago: University of Chicago Press.

Spring, A., C. Smith and F. Kayuni. 1983. **Women Farmers in Malawi; Their Contributions to Agriculture and Participation in Development Projects.** Chitedze, Malawi: Women in Agricultural Development Project. Mimeo.

PART IV

CASE STUDIES IN
LIVESTOCK PRODUCTION

9

Evaluation of Technological Alternatives for Small Farmers in Central America

Jeffrey R. Jones

Technology Development in Farming Systems Research

One of the most distinctive aspects of Farming Systems Research is its strategy for developing new technologies, which attempts to ensure on-farm applicability of technologies through an interaction of experiment station data and farm trial results. The first stages of technology development begin in the "characterization" phase of farming systems work, where farmers' problems, constraints and objectives are initially identified. Once alternative technologies have been identified a series of "validation" procedures are followed, either to verify their ultimate usefulness to farmers or to make necessary adjustments to the technology before making a generalized recommendation as to its use. Not all of these verification procedures are validation in a strict sense, but they occupy a homologous position in the research process (see Figure 1.1). Nor are these different validation procedures exclusive, but rather are complementary and usually sequential, to be applied according to the state of development of the alternative technology in question (Shaner **et.al.** 1982).

The focus of this paper is "technology evaluation"[1]. Technology evaluation is necessary when technologies have long production cycles which do not permit their installation and maturation within a project time frame. Projects which deal with perennial crops, forestry or animal production are the most likely to require this sort of analysis. The data presented here are taken from a technology evaluation of the CATIE-ROCAP Mixed Systems for Small Farmers Project[2], in Cariari, Costa Rica, and in Comayagua, Honduras[3]. The objective of the

Mixed Systems project was to develop technologies appropriate to the nearly universal "mixed" animal-crop production strategy practiced by small farmers in Central America. Although cattle and crops are complementary activities on small farms (see Muller 1982) they are usually treated separately in agricultural research (McDowell and Hildebrand 1980) with the result that grain crop production strategies are developed without reference to plant-animal interactions, and vice versa. The objective of the Mixed Systems project was to improve productivity per ha. for cattle and small animals by introducing improved dietary regimes, new forage crops, improved management practices, and at the same time improve the productivity of traditional crops[4]. The use of the technological alternatives required an increased use of capital and labor for their implementation, and it was hypothesized that there could be conflicting demands on these resources arising from the farmers' strategy for family and farm development and project technology requirements. The structure of farmers' goals was determined through the use of a methodology of "paired comparisons" (Harper and Eastman 1980), to analyze the changes in resource use implied by the adoption of the technological alternatives as outlined by project technicians.

Institutional Setting - Farming Systems Project Background at CATIE

In 1973, CATIE began with a broad project called the "Small Farmer Cropping Systems Project" (SFCS) in response to observations made with regard to agriculture in Central America. These observations are well summarized in a 1980 evaluation of the progress of the SFCS project;

> "... a survey of Central American agriculture ... had lead [researchers at CATIE] to the following conclusions: 1) most of the basic food staples, beans and corn in particular, were produced by small farmers whose average farm unit was less than 5 hectares; 2) most of the beans and corn produced were cultivated using multicropping rather than single cropping techniques; and 3) the agricultural technologies produced by international, regional or national research centers did not reach the small farmers who were using traditional, low-input technologies." (Hobgood **et. al.** 1980;2).

The SFCS project had addressed these problems, but the 1980 evaluation also found that the impact of the new research approach itself suffered from limitations. Although the cropping systems approach had investigated cropping strategies which were more immediately applicable to small farmers, it did not address questions of mixed crop-animal production strategies, despite the fact that this was a near universal production pattern. The "Mixed Systems for Small Farmers Project" began in 1981. This project was defined as a Farming Systems Research project, and while it built on the experience of the SFCS project in crops and the previous experience in other animal production projects (CATIE 1983, Novoa 1983a, 1983b, 1983c, 1984), it was directed toward a broader range of considerations.

Animals and crops are produced in a close symbiotic relationship on small farms; animals provide traction to plant or prepare land for cultivation. Animals consume crop residues, and successive crops are fertilized by animal wastes. In a managerial sense, capital, labor and land must be allocated for maintenance or production activities either for crops or for animals, so the relation between crops and animals is at the same time complementary and competitive. The importance of this relationship is demonstrated by the observation cited by Foster (1973;101), that farmers in Nepal rejected high yielding, dwarf grain varieties because they did not produce sufficient stalk residue to feed their animals. Cattle also serve an important financial function, in being a liquid store of capital; this function has been most elaborately described for tribal societies (see esp. Goldschmidt 1975 regarding African cattle keepers), but it is equally important in peasant farming societies (Muller 1982). The Mixed Systems project was designed to identify strategies which would permit improved animal **and** crop production through an analysis of complementary interactions for the design of production alternatives.

In 1983, the investigation described here was begun, with the objective of evaluating the acceptability of the technologies developed by the Mixed Systems project. On-farm implementations of technological alternatives were evaluated to permit a more realistic appreciation of their appropriateness through the observation of farmer reactions to and commentaries regarding on-farm trials. Although the focus of the Mixed Systems project was regional, at the time of the research reported here only the work at Cariari in Costa Rica, and Comayagua in Honduras was sufficiently advanced to permit an evaluation of the technologies in the field.

Project Area Background

The Central American Isthmus is divided into seven countries, Belize, Costa Rica, El Salvador, Guatemala, Honduras, Nicaragua and Panama. Its total population was more than 22 million in 1984, on a land area of some 516,000 km^2.

The area can be divided into three general climatic zones (see map). The most extensive climate area is that of lowland humid forest on the Atlantic side of the Isthmus, covered by dense broadleaf forest, and in some areas, open pine savannah. In general, precipitation ranges from 2,000 to 6,000 mm, with a short or no dry season. The area is low, generally below 600 m elevation with average annual temperatures greater than 20 C.

A second climatic zone is the highland area of the central mountain range which runs through most of the Isthmus. It is characterized in general by a relatively heavy rainfall, but with lower temperatures than in the Atlantic zone. Two major highland areas can be identified. A northern highlands area stretches from Mexico, through Guatemala, El Salvador and Honduras, and ends in northern Nicaragua. A southern highlands area runs from the north of Costa Rica to central Panama. Generally speaking, the northern highlands are drier than comparable areas in the south and have a longer dry season. They are characterized by pine forests alternated with smaller areas of broadleaf forest, while the southern highlands have only broadleaf forests.

A third zone, with a relatively extended dry season, is found on the Pacific side of the Isthmus, from the north of Costa Rica to the Guatemala-Mexico border. Smaller areas with similar climates can be found in the northeast of Guatemala, the north of Honduras and in parts of the Pacific coast of Panama. The marked dry season in these areas lasts three to eight months, with annual average precipitations usually between 1,000 and 1,500 mm. In terms of altitude and temperature, this zone is similar to the rainy Atlantic zone, although maximum temperatures may be higher.

The population of Central America is concentrated mainly in the highland areas of Guatemala, Honduras and Costa Rica, and in the pacific Lowlands of all countries. Agriculture as well is concentrated in the highland regions, and in the relatively dry Pacific regions. Except in ports and major banana-producing areas (which in most cases are contiguous), the humid Atlantic coast is sparsely populated.

The two project implementation areas discussed here are located in contrasting climate zones, although both are typically

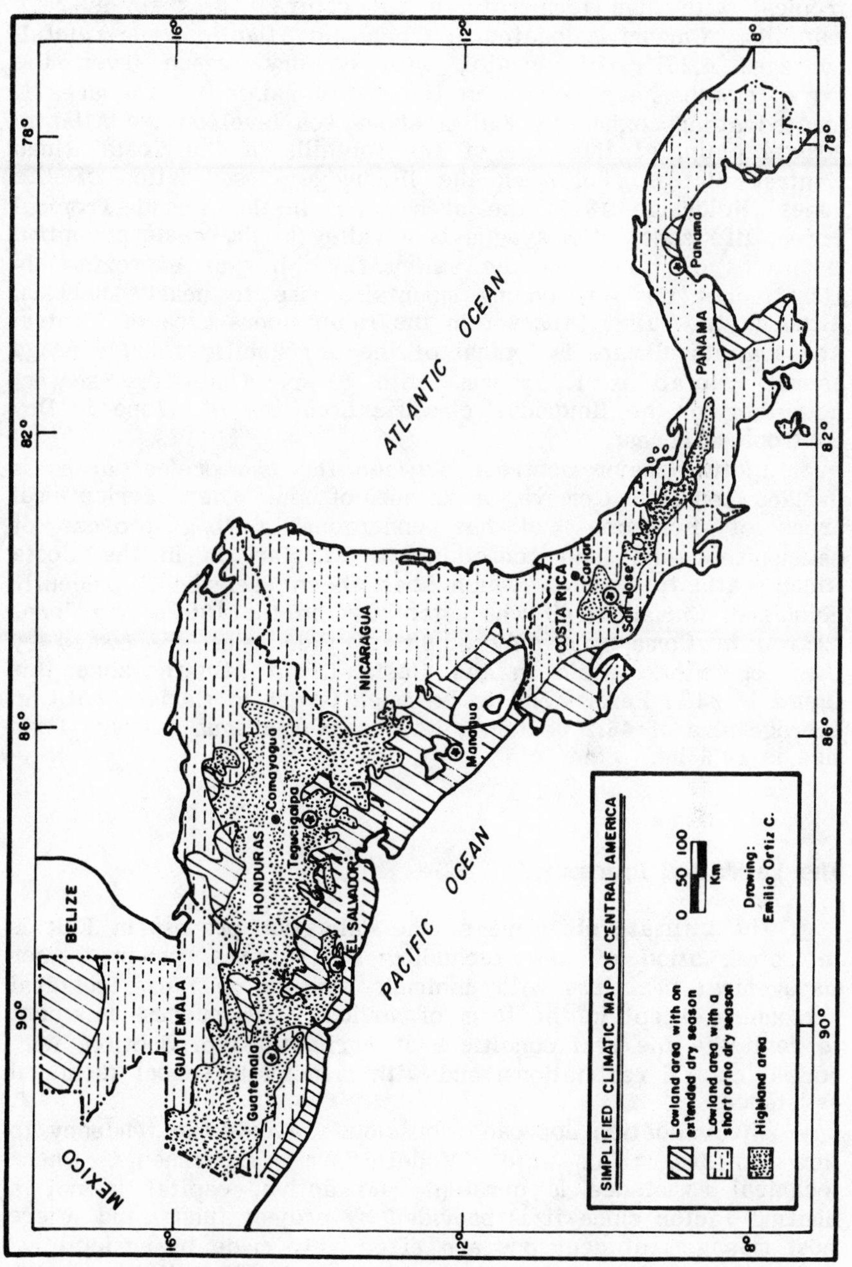
SIMPLIFIED CLIMATIC MAP OF CENTRAL AMERICA
Lowland area with an extended dry season
Lowland area with a short or no dry season
Highland area
0 50 100
Km
Drawing: Emilio Ortiz C.
ATLANTIC OCEAN
PACIFIC OCEAN
MEXICO
BELIZE
GUATEMALA
HONDURAS
EL SALVADOR
NICARAGUA
COSTA RICA
PANAMA
Guatemala
Tegucigalpa
Comayagua
Managua
San José
Cariari
Panamá
16°
12°
8°
78°
82°
86°
90°

tropical with high temperatures but contrast in their average rainfalls. Cariari is located in the humid Atlantic zone; rainfall averages 4,261 mm annually, with no dry season (even the dryest months have more than 100 mm of rainfall). The area is located at approximately 300 m above sea level on the Atlantic coastal plain, at the base of the foothills of the Costa Rican Central Plateau. Following the Holdridge classification of life zones (Holdridge 1979) the area falls in the Humid Tropical Forest life zone. Comayagua is a valley in the western central mountains of Honduras; the valley floor lies at approximately 1000 m but the surrounding mountains rise to nearly 3000 m. Although the valley falls within the mountainous area of Central America, its climate is typical of the dry Pacific zone; average annual rainfall is 1,035 mm, with a six month dry season. According to the Holdridge classification, the life zone is Dry Subtropical Forest.

Another major contrast between the two project areas is in land tenure. Comayagua is one of the oldest agricultural areas of Honduras and has undergone a long process of development and land consolidation. Cariari is in the Costa Rican Atlantic zone, which has been relatively recently colonized; Cariari itself was established in the 1960's (see Jones 1985a). In Comayagua only 32.6% of the farms are owned by their operators, while in the Costa Rican Atlantic zone this figure is 93%. Farms tend to be much larger in Cariari, with an average size of 46.2 hectares, while in Comayagua average farm size is 10.9 ha.

The Validation Process

The ultimate objective of the validation process in FSR is the observation of new technologies on farms, under farmer management practices with minimal intervention from technical personnel, except in the form of advice. This process is meant to replicate the real conditions of agricultural production, with normal capital restrictions, and with principally verbal technical assistance.

The validation approach consciously avoids the tendency to turn on-farm trials into "Model Farm" experiments, where technical assistance is constant and active, capital is not a limiting factor since it is provided by project funds, and where most management decisions are effectively made by agricultural technicians rather than by the farmers. The pitfall of the "Model Farm" approach should be obvious: a few hundred dollars injected into a poor farm can completely change the income horizon for the farmer, to the extent that new income maximization strategies involving project manipulation are a major factor in on-farm decision making: or, the farm can be run as an experiment station, with a level of investment of capital, technical assistance and other resources which are not normally available on farms. In both cases, it is possible to produce very good agronomic results which have little relation to local socio-economic conditions. It is likely that the "Model Farm" phenomenon is the cause for the disappearance of apparently successful agricultural technologies when project funds are removed.

The validation of technology requires several complete reproductive cycles. Once a technological alternative is developed it is introduced to farmers for management on their own farms. At this stage, farmers may receive project funds to ensure that they have sufficient resources to properly implement the alternative. It is assumed that the technology has been sufficiently tested in experiment station trials for there to be little doubt as to its technical efficacy, and that in the initial year of use there will be a substantial income improvement which will provide funds for the coming season. After a period of instruction, farmers are "separated" from the project in the sense that they cease to receive whatever support or subsidies were necessary for the initial period of demonstration. An "index of acceptability" (Hildebrand 1979) can be calculated on the basis of the number of farmers who continue to use the technology once project financial support is removed, since this should indicate the percentage of farmers who, knowing how to use a technology, find it worthwhile.

Validation exemplifies the interactive character of FSR, in that it is designed to allow farmers to concretely demonstrate the viability or problems of technological alternatives under normal socio-economic conditions. Observations by both agricultural technicians and the farmers are reincorporated into the technology design process. Of particular interest are spontaneous adjustments farmers make to the proposed technology during the validation period in response to problems encountered in the field.

In certain cases, it may be desirable to carry out on-farm trials even though the alternative technology is not entirely ready for validation. This occurs when a complex or slowly reproducing system is implemented. For the Mixed Systems technological alternatives several seasons are necessary to establish fodder crops, reproduce animals and make necessary installations of the farms. The efficiency of these systems depends on the interaction between different components, which can only be appreciated once the entire system is established and functioning. In such cases, the technology must be "evaluated", rather than "validated" using methods which try to gauge technological utility and compatibility on the basis of indirect measures, e.g. comments by farmers, or observations of their on farm resources and estimations of farmer capabilities.

Procedures for Technology Evaluation

Technology evaluation faces a special set of problems since results must be projected onto a larger population on the basis of interviews and observation of farm activities on few farms. For the purposes of evaluation, it is imperative that farmers critically analyze and comment on the new technologies from the perspective of their applicability under **their own** farm conditions, as a simulation of actual validation.

A first problem in the evaluation is to establish that farmers understand the technology being presented. Since the use of a technology implies the commitment of resources, possible conflicts with other activities on the farm, and changes in production procedures, the farmer cannot evaluate its applicability without a thorough understanding of what the new technology entails.

A second problem is; how to ensure that farmers really express their critical observation of technologies? It is common for farmers to see experiments in much the same way as they see their own farms, as extensions of their personalities, work capacity, responsibility, intelligence and management ability. A strong sense of etiquette is implied in any observation of another's agricultural activity, and the first reaction of the farmer is most likely to be one of polite reserve; you would no more criticize an acquaintance's farming than you would criticize their clothes, in "proper" daily interactions.

Finally, an understanding of the farmers' motivational framework must be developed to permit a comparison of the farmers' aspirations and the requirements for the implementation of the technology.

These problems were resolved through the use of interviews with farmers carried out in multiple visits to their farms. In an initial visit, the purpose of the project and evaluation exercise were explained, and farmer's goals for their family development were established. In a second visit, farm development goals were established, and their participation in a visit to the demonstration farm was confirmed. The visit to the demonstration farm was followed by still another interview, to collect farmer's observations with regard to the applicability of the alternative technology on their own farms.

A more complete understanding of the alternative technology was thus promoted in several ways. During the initial interviews, the problems addressed by the alternative technology were discussed with the farmers, and the solutions proposed by the alternative technology were presented verbally. The strategy was to focus the farmer's attention on the technology, if it was not already, and to let him contemplate strategic alternatives during the period between interviews, so that during the visit to the demonstration farm he would be more prepared to critically observe. In the preliminary talks, questions were deliberately posed and left open, to be resolved in the visit to the demonstration farm.

The visit to the demonstration farm involved an inspection of the entire farm, a guided tour with the owner as guide. Technicians accompanied the group, but only responded to direct questions which were posed to them by group members; it was to be clear that the farmer was in charge. The farm tour was loose and unstructured, with a continual question and answer process. At the end of the visit, a final group discussion was held for clarifying final questions; at this point the technicians made a brief presentation as to how and why they thought the system should operate. It was felt this was necessary, because of misconceptions which may have been exchanged among the farmers during the visit. Since the project was still in progress, the educational element was felt to be important.

Several elements were included in the evaluation process to ensure that farmers would be able and disposed to make critical evaluations of the alternative technology. First, during the initial interviews an attempt was made to establish who besides the farmer was involved in farm decision making. In many cases it was the wife, although in some it was a brother, son or friend; the farmer was encouraged to attend the demonstration with that person. The final observations by the farmer on the alternative technology were recorded at his own farm, several days to a week after the visit to the demonstration farm. This allowed time for the alternative to be discussed, and also ensured that the farmer was not put in the ungracious position of being asked to "criticize" his "host's" farm during what was for many a social visit. In addition, an attempt was made to discuss the application of the technology in the context of the stated goals from the earlier interviews, and determine if these goals would be more easily achieved through its use. During all the interviews, it was stressed that we, as project technicians, were involved in an experiment, and that we honestly did not know enough of the details of farming in the area to completely evaluate the applicability of the technology.

Due to the intensiveness of the multi-visit strategy, a limited number of farmers could be included in the evaluation process. In Cariari, 12 farmers were included, and in Comayagua there were 18 (more were interviewed, but some had incomplete information due to time constraints). These samples constituted 3-4% of the population. Since this was felt to be a sufficiently large sample statistically, extra effort was dedicated to guaranteeing the quality of the information gathered.

Since interview time was to be used in the collection of attitudinal data, families included in the samples were limited to those for which up-to-date socio-economic information was available. This placed obvious constraints on sample selection. However, in both Cariari and Comayagua sufficient data had been previously collected for some 80 randomly selected farms (in each area), and it was possible to select subsamples which were structurally similar to the target populations in terms of land and animal possession.

Comparison of Farming Systems

Farming Systems of the two project areas are quite different. Comayagua agriculture is focused around production of corn and sorghum on unirrigated lands and vegetable production on irrigated land. Cariari farms focus on the production of corn and tubers. Both areas share a general small farm orientation toward the management of increasing numbers of cattle, which serve a "banking" function; the value of the animals purchased grows with a minimum of attention, and they can be quickly turned into cash when necessary, or used as collateral for loans.

In Comayagua, farms are developed through the progressive purchase of small parcels. Young families first establish themselves on a houselot in a village or town, and begin farming inherited or rented parcels. If land is irrigated, high value vegetable crops can be grown, such as watermelon, tomato, green peppers, etc. For capital poor farmers, cash inputs can be "financed" through sharecropping arrangements, since wealthier individuals in some cases prefer letting others manage these labor intensive crops[5]. A typical small farm will have 3 to 7 parcels averaging 1 ha. in size, some irrigated and some not irrigated. Income is derived primarily from vegetable sales, and additionally from subsistence grains when there is surplus production. In good years, farmers purchase cattle with their profits, and these are rotated along with the oxen to different parcels to feed on crop residues.

The critical aspect of the farming operation is the maintenance of the farmers' "capital", the cattle, through the dry season. Farmers try to sow sufficient forage for their animals, either in the form of the stalks of corn and sorghum planted for grain, or through the high density planting of sorghum, exclusively for animal feed. Cattle can also be maintained in public lands, although this is an unpredictable resource of generally poor quality. The length of the dry season varies, and in years where the dry season is longer than expected, farmers must resort to renting pastures, buying feed or selling animals. Some farmers purchase infertile, sloping land in or near the valley as a dry season pasturage for their animals. The difficult management decision in this case is how much land, labor and capital to invest in forage production, at the cost of cash crop production.

Cariari is much more heterogeneous in the farming systems observed. In terms of size, farms tend to be more homogeneous than in Comayagua; approximately half of the farms are still in the hands of the original land reform beneficiaries, and the rest have generally changed hands as the whole 20 ha blocks initially distributed to each farmer, rather than be subdivided. Corn and cattle production are the major sources of cash income, but a great number of additional crops of less importance, or which are in the experimentation stage, are also grown. Cassava is an important cash crop in the area, but this depends on variable international markets. New specialty crops such as ginger, curry, black pepper, palm heart, tropical fruits, etc. are found on many farms, but they are not uniformly distributed and are not yet considered a permanent part of most farms.

A general pattern for farm development in Cariari is a focus on high return, labor intensive and risky activities in order to establish working capital on the farm. These crops are most commonly cassava and corn; commercial pig production also falls into this category but to a lesser degree, as do some of the specialty crops mentioned above. As a farmer builds up capital he buys cattle, but with more of a "production" (as opposed to "banking") emphasis than in Honduras. Since farms are relatively large, farmers can expect to earn a reasonable living from cattle reproduction, although they may focus on the sale of calves to larger farms for fattening. Nevertheless, cattle production requires a relatively large farm to allow a farmer to generate an attractive income, so the cattle option requires the purchase of additional farmland to become a long-term option for farmers.

In the farming systems of both areas, the use of extensive

cattle grazing was an integral part of the farmers' personal development strategy. The objective set for the Mixed Systems Project was the introduction of cattle management techniques to permit improved income through the use of cultivated pastures and forages, which would result in better animal health without resorting to land extensive grazing, lower mortality rates and shorter periods between births. These activities were directed especially at the non-irrigated lands in Comayagua, since these were those most misused and least productive lands in the area. In Cariari, project activities were oriented toward increasing the protein content of animal feeds, since in high rainfall areas animals tend to be undernourished due to the intake of low quality feeds. In both Comayagua and Cariari, the project proposed to work with animals which are genetically "superior" to those commonly used by the farmers (milk cows in both areas, and pigs and cows in Cariari).

Briefly described, the technical improvements for Comayagua were;

1. the plantation of sugarcane for use as feed during the dry season.
2. the plantation of **Leucaena** spp. as a protein source, and as a method to overcome food shortage in dry months, since its deep tap root would reach deeper water supplies than grasses or annual crops.
3. changes in fertilization and plant spacing in corn production, to permit a maintenance of maize production on a reduced land area.
4. the use of Zebu cattle stock with slight mixtures of Holstein or Brown Swiss to improve milk production capacity.
5. the use of the forage chopping machine to permit the proper mixtures of feeds, and the use of otherwise unpalatable food sources.

For Cariari, the improvements proposed were;

1. the introduction of a leguminous cover crop, which would improve soil quality, and produce high protein feed for cows and pigs.
2. the use of cassava residues to improve cattle feed.
3. the plantation of King Grass (**Pennisetum** sp.) as a mainstay for the milk cows' diet, to reduce the need for pasture area.
4. the use of a machine to chop forages and allow the feeding of a proper mixture of different feeds to cattle.
5. introduction of Jersey cattle, to minimize food and

especially protein needs.

6. increased use of native **criollo** pigs, rather than crosses with foreign pigs, to reduce protein requirements.

Determination of Farmer Goal Preferences

Lists of alternative goals were developed with reference to the farm family and to the farm as a production unit. The lists were deliberately kept short to avoid confusion in the collection of data (See Urquhart and Eastman 1978).

Alternative goals for **family** development were:

1. Self sufficiency in food production.
2. Improvements in the quality of life through the acquisition of material possessions.
3. Prestige within the community (respect of neighbors).
4. Children's education.
5. The assembly of a material inheritance for children (land, houses, money, etc.).
6. Leisure (activities not directly related to farm production).

Alternatives for **farm** development were:

1. Acquire more land.
2. Acquire more on-farm capital.
3. Increase income.
4. Assure a constant income.
5. Avoid risk (especially bank loans requiring a farm mortgage).
6. Diversify production.
7. Avoid hiring off-farm laborers.

Each of the lists of goals were presented in the course of an entire interview (two interviews were required). At least one hour was spent discussing the definitions of goals before actually ranking them. For example, the farmer was asked what was the level of education to which he aspired for his children, where they would have to go to receive it, etc. as a prelude to ranking the "education" goal with the other alternative.

The ranking of alternative goals was done through a method of paired comparisons (Harper and Eastman 1980; Urquhart and Eastman 1978). This method was chosen because it permitted ambiguity in the ranking[6]. During the course of the interviews the advantages of the method were apparent. Farmers frequently commented that they found the comparisons and rankings of preferences difficult, because they in fact tried to structure their lives so as to avoid the necessity of choosing. They commented in some instances that there was no real need for them to choose between certain goal alternatives, for example, "because school board meetings (education goal) are scheduled on other days than other community meetings (prestige goal)". It seems likely that on a collective level an effort was made to avoid these conflicts. In all cases the interviewer initially insisted on a clear statement of priority, although a tie was recorded if the farmer would not state a preference.

It was especially important to do this type of interview with a minimum of peer interference. While the presence of children did not affect interview results, the appearance of neighbors visibly altered the interview situation, by distracting the farmers' attention, on the one hand, and due to the opinions and suggestions offered by the neighbor in the course of the interview.

Presentation of Technological Requirements for Technological Alternatives

The presentation of the technological requirements of the technological alternatives involved considerations of labor inputs, costs, profitability, and general desirability of the proposed activity. The comparison of farmer goals and technological requirements was carried out in the final interview after the visit to the demonstration farm. To prepare the interview, the technological alternatives were considered from the point of view of required inputs, and the difficulty of acquiring them, and questions were directed toward these areas.

During the interview, specific questions were addressed to the scale of operations, labor requirements and the need for on farm labor reallocation. While these questions were presented in a standard format, interviewees were encouraged to expand on each of these topics, and especially on aspects which they found especially appealing, or especially dubious with regard to their applicability.

With regard to costs, the discussion was focused through the use of a simplified cost-income chart which demonstrated project estimations and assumptions.

Analysis of Paired Comparisons

The same set of goals was tested in both Cariari and Comayagua, with the adjustment of terminology in the field. The samples were broken down into two classes in each study area, since cattle are used as liquid capital investments, their number was operationalized as an indication of wealth. Farm development goals and family development goals were analyzed for each "wealth" class, "wealthy" individuals being those with more than 20 adult animals.

The preference ordering for family development goals showed a great homogeneity, and was virtually identical for both wealth classes in both areas. The education of children was universally cited as the most important family goal, while leisure time was the least important goal. In Cariari, the ordering was not differentiated by strong preferences; in Comayagua, preferences were more differentiated (See Table 9.1) with education singled out as a unique, significantly preferred goal. Even when the significance level of the LSD test is reduced to a probability of less than .20, Cariari farmers do not show the degree of preference for education seen in Comayagua.

In the preference orderings for farm development goals, there is less homogeneity than in family development goals. In each of the study areas, both classes of farms agreed on the least important goals; in Cariari, land acquisition was felt to be the least important goal in agricultural production, while in Honduras, the avoidance of contracting outside laborers was felt to be the least important goal. The avoidance of risk was a more important goal for the wealthier farmers in both Comayagua and in Cariari, although only in Comayagua does it reach a point of statistical differentiation from the other goals at a .10 level (See Table 9.2).

Interpretation of Paired Comparison Data

Family development goals for Cariari and Comayagua are differentiated by the strong preference for education as a goal for Honduras and a relatively weak preference in Cariari. The differences in the degree of preference for the goal of education of children reflects the educational situation in the two countries. Costa Rica has an extremely effective national education program, and high degree of literacy, when compared to Honduras. Nearly all Costa Ricans finish the first 6 years of instruction, and it is not uncommon for a full 12 years to be

Table 9.1
Family development goals in Cariari, Costa Rica
and Comayagua, Honduras

Cariari - Family Development Goals by Number of Cattle

0 - 20 Animals		20 - 80 Animals	
LSD P LT* .05 = 12.70		LSD P LT .05 = 12.70	
LSD P LT .10 = 10.96		LSD P LT .10 = 10.96	
LSD P LT .20 = 8.31		LSD P LT .05 = 8.31	
Goal/	0 - 20 Animals		20 - 80 Animals
Education	19	Education	20
Self Sufficiency	15	Self Sufficiency	20
Inheritance	15	Inhertance	15
Quality of Life	13	Quality of Life	12
Prestige	13	Prestige	11
Leisure	9	Leisure	3

Comayagua - Family Development Goals by Number of Cattle

0 - 20 Animals		20 - 80 Animals	
LSD P LT .05 = 17.20		LSD P LT .05 = 17.20	
LSD P LT .10 = 14.43		LSD P LT .10 = 14.43	
LSD P LT .20 = 10.36		LSD P LT .20 = 10.36	
Goal/	0 - 20 Animals	Goal/	20 - 80 Animals
Education	41	Education	48
Self Sufficiency	40	Inheritance	35
Quality of Life	29	Self Sufficiency	24
Inheritance	31	Quality of Life	21
Prestige	18	Prestige	18
Leisure	5	Leisure	18

All Farms - Family Development Goal Structure

Cariari		Comayagua	
LSD P LT .05 = 17.96		LSD P LT .05 = 24.32	
LSD P LT .10 = 15.08		LSD P LT .10 = 20.41	
LSD P LT .20 = 11.75		LSD P LT .20 = 15.91	
Education	39	Education	89
Self Sufficiency	35	Inheritance	64
Inheritance	30	Self Sufficiency	64
Quality of life	25	Quality of Life	52
Prestige	24	Prestige	36
Leisure	12	Leisure	23

* "Probability Less Than ...". LSD is the "Least Significant Difference" between two scores at a given level of significance. The vertical lines group variables which are not significantly differentiated at the .05 level.

Table 9.2
Farm Development Goals in Cariari, Costa Rica
and Comayagua, Honduras

Cariari - Farm Development Goals by Number of Cattle

LSD P LT .05 = 14.67
LSD P LT .10 = 12.31
LSD P LT .20 = 9.59

LSD P LT .05 = 14.67
LSD P LT .10 = 12.31
LSD P LT .20 = 9.59

Goal/	0 - 20 Animals		20 - 80 Animals
Constant Income	27	Avoid Risk	28
Increase Income	22	Constant Income	27
Avoid Risk	21	Increase Income	21
Diversify Farm	18	Avoid Laborers	19
Increase Farm Capital	15	Increase Capital	17
Avoid Hiring Laborers	13	Diversify Farm	17
Increase Farm Size	11	Increase Farm Size	6

Comayagua - Farm Development Goals by Number of Cattle

LSD P LT .05 = 15.84
LSD P LT .10 = 13.30
LSD P LT .20 = 10.36

LSD P LT .05 = 18.94
LSD P LT .10 = 15.89
LSD P LT .20 = 12.39

Goal/	0 - 20 Animals	Goal/	20 - 80 Animals
Increase Income	26	Avoid Risk	51
Diversify Farm	25	Constant Income	34
Constant Income	24	Increase Farm Capital	33
Avoid Risk	23	Increase Income	33
Increase Farm Size	22	Diversify Farm	32
Increase Farm Capital	21	Increase Farm Size	17
Avoid Hiring Laborers	6	Avoid Hiring Laborers	5

All Farms - Farm Development Goal Structure

Cariari
LSD P LT .05 = 20.74
LSD P LT .10 = 17.41
LSD P LT .20 = 13.57

Comayagua
LSD P LT .05 = 24.69
LSD P LT .10 = 20.72
LSD P LT .20 = 16.15

Cariari		Comayagua	
Constant Income	54	Avoid Risk	74
Avoid Risk	49	Increase Income	59
Increase Income	43	Constant Income	58
Diversify Farm	35	Diversify Farm	57
Increase Capital	32	Increase Capital	54
Avoid Laborers	32	Increase Size	39
Increase Size	17	Avoid Laborers	11

completed. Agricultural "colleges" (agriculturally oriented secondary schools) are highly dispersed in rural communities. In contrast, the educational level in Honduras is much lower; many children do not attend school, and illiteracy is high. Conversely, the perception of job opportunities and improved income through educational achievements is much higher in Honduras, probably accurately reflecting the unfilled demand for educated workers. In Costa Rica, the phenomenon of unemployed professionals has grown; several farmers made the observation that they personally knew of unemployed, relatively highly educated individuals, and concluded that their children may well have better opportunities in agriculture.

Another important factor is the observable quality of life in the agricultural sector. Cariari is a government sponsored colonization project which enjoys a high degree of government support in the form of technical assistance, bank credits, and infrastructural development. Comayagua farmers, on the other hand, generally do not receive as much government support, and the agricultural opportunities are visibly reduced due to problems of climate and the lack of infrastructure. Costa Rica traditionally has been a country which depends in large degree on the production of small farmers, even for export production (especially coffee), and this is reflected in the quality of government support which smaller farmers receive. A small farmer in Cariari seems much more likely to achieve an acceptable quality of life than his homologue in Comayagua, so the Honduran farmers tend to look outside the agricultural sector in their plans for their family.

Farm development goals show more consistency for wealth classes than for their location. There is a marked economic conservatism among the wealthier farmers of both areas, while the poorer farmers seem more disposed to risk or select otherwise undesirable alternatives in hopes of increasing income. In Cariari, it was universally observed that the contracting of laborers was difficult and unpleasant, due to the scarcity of labor and the procedures for satisfying labor laws (especially paying insurance, retirement and disability allowances). Nevertheless, poorer farmers are less averse to contracting labor in order to increase income or farm capital (P less than .10). In Comayagua, wealthier farmers are much more averse to risk (in the form of bank loans) than are poorer farmers (P less than .10), and were not disposed to borrow money for any purpose, while poorer farmers were not so categorically averse to loans for farm or income improvements.

Due to the relatively small sample sizes for the two areas (12 farms in Cariari - approximately 4% of the target population - 18 farms in Comayagua - 3.5% of the target population) the preference structures of the farmers were not highly differentiated at the P less than .05 level, and the P less than .10 level was used as a basis for comparison. The interpretations based on these analyses are presented because they largely corroborate more intuitive observations made by investigators in the area. Obviously, a definitive study would require larger samples and more stringent decision criteria, but for the purposes of the Mixed Systems Project (and obviously, within the limits imposed by time and budget) it was felt that these data were sufficiently clear to make project recommendations.

Conclusions and Recommendations

Two basic conclusions, and recommendations, were derived from the analyses of farmer goals. First, poorer farmers were seen to be more receptive to changes which required sacrifices on their part. This conclusion in some degree contradicted the working assumptions of the project, since it was assumed that due to their poverty, poorer farmers would have less capital and time to invest in new activities which did not contribute primarily to food production. However, the results indicate that wealthier farmers are less willing to risk their accumulated capital, or to involve themselves in undesirable activities (such as labor contracting to make up for family labor deficits in intensified cultivation regimes) in hopes of achieving higher incomes. It was recommended that project activities give more attention to poorer farmers, and that stepwise implementation techniques for the alternative technologies be investigated to make them more accessible to capital poor farmers.

Second, the ultimate objective of the Mixed Systems Project was the formulation of technical packages whose increased capital intensity could be offset by increased profits and financing by local banks. As observed above, bank loans were not well received by farmers in general, although some indicated a willingness to use these funds in relatively risk-free activities. However, in Honduras this strategy was clearly not feasible; farmers' determination to educate children resulted in many cases in a decapitalization of the farm to pay educational expenses. The clear preference for education of children, and the determination that children should not work in the agricultural sector meant that capital investment in farms, even if it increased the sale value of the farms (and presumably the value of the inheritance for children) was not a desirable alternative. As a result, it was recommended that efforts be made to decrease the need for investment in equipment in the Honduran technological alternatives.

As a methodological conclusion, it should be clear that paired comparisons can be applied to ranking of preferences other than the goals selected for the Mixed Systems evaluation. The selection of goal alternatives to be ranked will depend on the formulation of hypotheses to "explain" potential non-acceptance. For the evaluation described here, the hypotheses were that labor and capital requirements of the alternative technology could conflict with farmers' goals. Specific hypotheses must be developed for each study on the basis of existing conditions and problems encountered with the technology to be evaluated.

Data collection and analysis reflects the multidisciplinary exigencies of FSR field work. A conscious attempt was made to merge qualitative anthropological field investigation techniques and more quantitative statistical analysis. Interviews were lengthened, and the form of the questions was altered to permit a more "ethnographic" data collection procedure, which would avoid the problems associated with the rapid application of standardized interview schedules. The "paired comparison" technique provided a degree of statistical proof which helped biological scientists understand how conclusions were derived.

In the process of technology evaluation, special emphasis must be placed on ensuring that the farmer understands the motivations for proposing the technological alternative, and the framework within which he is being consulted. Reasons of courtesy, timidity, or insufficient understanding of the evaluation can contribute to the farmers being reticent to express their observations with regard to the technologies in question. Since the effectiveness of the evaluation depends on success in enlisting the farmers' intelligent participation in the process, it is justifiable to dedicate more time and effort to guaranteeing their understanding of the project.

Notes

1. Some parts of this paper were presented in the Short Course **Agroforestry for the Humid Tropics**, CATIE, Turrialba, Costa Rica, April 24 - May 4 1984.
2. ROCAP is the Regional Office for Central American Programs of the US Agency for International Development. CATIE is the Centro Agronomico Tropical de Investigacion

y Ensenanza, based in Turrialba, Costa Rica.

3. Data are the product of 2 months of field work. Arturo Vargas assisted in data collection in Cariari, as did Edwin Cruz in Comayagua. Marcelino Avila, Carlos Burgos and Marco Esnaola were constant sources of support and ideas in this activity at CATIE in Turrialba, Costa Rica. Medardo Lazo, Enrique La Hoz and Roger Meneses directed field activities for the project in the field, and were extremely helpful in identifying potential goal and resource conflicts to be investigated.
4. The technologies are more completely described in Jones 1983a, and Jones 1983b.
5. Sharecropping as a form of financing for expensive crops is discussed in more detail for Bolivian potato farmers in Jones (1985b).
6. The analysis of the paired comparisons is carried out by summing the number of times a goal is preferred to other goals. The totals are summed for all cases for each goal. The numerical difference between these scores are then compared by means of a calculation of the "Least Significant Difference" (LSD), which is calculated as: LSD $= 1.96\ [B(t)(t+1)/6]^{\frac{1}{2}}$, where "B" is the number of interviews and "t" is the number of objectives which are included in the analysis.

Bibliography

CATIE (Centro Agronomico Tropical de Investigacion y Ensenanza). 1983. **Investigacion Aplicada en Sistemas de Produccion de Leche: Informe Tecnico Final del Proyecto CATIE-BID 1979-1983.** Turrialba, Costa Rica: CATIE.

Foster 1973. **Traditional Societies and Technological Change.** New York: Harper and Row.

Goldschmidt, Walter R. 1975. "A National Livestock Bank: An Instutional Device for Rationalizing the Economy of Tribal Pastoralists". In **International Development Review** 2; 2-6.

Harper, W.M. and Eastman, C.E. 1980. "An Evaluation of Goal Hierarchies for Small Farm Operators". In **American Journal of Agricultural Economics** Vol. 65 # 4; 742-747.

Hildebrand, Peter E. 1979.
Incorporating the Social Sciences into Agricultural Research: The Formation of a National Farm Systems Research Institute. New York: ICTA (Guatemala) and The Rockefeller Foundation.

Hobgood, Harlan, Rufo Bazan, Rollo Ehrich, Francisco Escobar, Twig Johnson and Mark Lindenberg. 1980. **Central America: Small Farmer Cropping Systems.** Project Impact Evaluation No. 14. Turrialba, Costa Rica: CATIE-ROCAP.

Holdridge, Leslie R. 1979.
Ecologia. San Jose, Costa Rica: Instituto Interamericano de Cooperacion Agricola.

Jones, Jeffrey R. 1983a.
Resultados del Trabajo de Pre-Validacion en Cariari, Costa Rica. Turrialba, Costa Rica: CATIE.

——. 1983b. **Resultados del Trabajo de Pre-Validacion en el Valle de Comayagua, Honduras.** Turrialba, Costa Rica: CATIE.

——. 1985a. **Land Colonization in Central America.** Final Report of Land Colonization Assessment in Humid Tropical Areas. Tokyo: United Nations University.

——. 1985b. "El Papel de los Intermediarios en la Produccion de Papa en Cochabamba, Bolivia: Aspectos Financieros de su Participacion en la Produccion". In **Comercializacion Interna de los Alimentos en America Latina: Problemas, Productos y Politicas.** G.J. Scott and M.G. Costello (eds.). Ottawa: IDRC.

McDowell, R.E. and Hildebrand, P. E. 1980.
Integrated Crop and Animal Production: Making the Most of Resources Available to Small Farms in Developing Countries. New York: The Rockefeller Foundation.

Muller, Eduardo. 1982.
"Cash-Crop with Animal Production System: Coffee, Sugarcane with Dual Purpose Cattle." In **Proceedings of a Workshop: Research on Crop-Animal Systems. Case Studies.** H.A. Fitzhugh, R.D. Hart, R.A. Moreno, P.O. Osuji, M.E. Ruiz and L. Singh (eds.). CATIE-CARDI-WINROCK. Morrilton, Arkansas: Winrock International.

Novoa, Andres R. 1983a.
Aspectos Nutricionales en la Produccion de Leche: Compilacion de Documentos Presentatos en Actividades de Capacitacion. Vol 1. Turrialba, Costa Rica: CATIE.

——. 1983b. **Caracterizacion y Evaluacion de Sistemas de Fincas en Produccion de Leche: Compilacion de Documentos Presentados en Actividades de Capacitacion.** Vol. 2. Turrialba, Costa Rica: CATIE.

——. 1983c. **Diagnostico de los Esquemas Institucionales para la Comunicacion y Transferencia de Tecnologia Agropecuaria en el Istmo Centroamericano.** Turrialba, Costa Rica: CATIE.

——. 1984. **Salud, Manejo y Administracion en Sistemas de Produccion de Leche: Compilacion de Documentos Presentados en Actividades de Capacitacion.** Vol 4. Turrialba, Costa Rica: CATIE.

Shaner, W. W., Philipp, P.F. and Schmehl, W.R. 1982.
Farming Systems Research and Development: Guidelines for Developing Countries. Boulder, Colorado: Westview Press.

Urquhart, N. Scott and Eastman, Clyde. 1978.
"Ranking Energy Policy Alternatives in a Partially Informed Population". In **Department of Energy Statistical Symposium, Albuquerque, New Mexico.** Oak Ridge, Tennessee: Oak Ridge National Laboratory; 53-65.

10

Adaptations of Farming Systems Research to the Study of Pastoral Production Systems: The Niger Range and Livestock Project[1]

John J. Curry, Jr.

Introduction

Until recently, a serious shortcoming in Farming Systems Research has been its preoccupation with the cropping aspect of the domestic production system, to the neglect of animal husbandry. Such an oversight on the part of most farming systems researchers prompted McDowell and Hildebrand (1980;5) to remark that;

> ..virtually none of the work that has been done has included the animal component, an important element of most of the world's small farms.

Recent work, including several of the cases presented in this volume, has endeavored to correct this oversight on the part of researchers. Studies in West Africa on crop/livestock interactions in mixed-farm and agropastoral settings by such organizations as the Center for Research on Economic Development at the University of Michigan (CRED) (Shapiro 1979) and ILCA (1984), for example, have successfully used a systems approach to fashion policy recommendations in the areas of livestock production and marketing. By focusing on the **total** domestic economy, rather than a single aspect of production, such approaches can eliminate an analytical bias which favors the cropping system to the neglect of livestock production.

Such an enhancement of the farming systems approach

should be well-suited, moreover, to policy-oriented studies of pastoral production systems, where livestock play a dominant role in local subsistence strategies, and social life. The integration of traditional management and ownership patterns into development strategies poses an important contribution to the development of areas with extreme environmental conditions (Goldschmidt 1975).

The Niger Range and Livestock Project (NRL) provides an example of a conscious attempt by policy-makers to adapt a farming systems style of research to a pastoral setting in order to fashion viable policy recommendations. From 1980 to 1982, the project utilized a modification of the "cost route" type of methodology often used in farming systems research to conduct an intensive investigation of domestic production among pastoral nomads and sedentary farmers in the Tahoua **department** located in the central region of the Niger Republic in West Africa. The socioeconomic research of the project was undertaken by a team composed primarily of social and economic anthropologists, who were responsible for the collection and analysis of the survey data.

The discussion which follows includes a brief introduction to the geography and ecology of the project area, a description of the project itself, some of the production systems — both agricultural and pastoral — which were studied by the project, and the contributions of the anthropologists who worked as members of the socioeconomic team. As a member of the project, I was responsible for the study of the inter-relationships between animal husbandry and agriculture among sedentary farmers and their pastoral and agropastoral neighbors in the southwest portion of the project area. The site chosen for this research was the village of Shadawanka, an important livestock market. Specific information on Shadawanka's farming system is derived from eighteen months' field study in the village, and among the surrounding hamlets.

The Niger Range and Livestock Project affords an opportunity to examine the contributions social scientists make to policy-oriented research when employed as integral members of Farming Systems Research teams.

The Niger Republic: Background for Pastoral Development

The Republic of Niger (Figure 10.1), which occupies approximately 1,267,000 km², is a landlocked country in the West African Sahel. As such, it shares borders with six other

West African nation states; Algeria and Libya to the north, Chad to the east, Nigeria to the south, Benin and Bourkina Fasso to the southwest, and Mali to the west.

Rainfall is perhaps the single most important determinant of habitat conditions in this portion of West Africa. The climate is characterized by a single, short rainy season (June to September), followed by a long, dry season (October to May). The rainfall regime is subject to a high degree of spatial and interannual variability. There are three major climatic/vegetational zones for Niger;

1. the Sudan Savannah Zone - a treed savannah ranging from the southern borders with Benin and Nigeria to 15 degrees North latitude that receives between 600 and 800 mm of rainfall annually.
2. the Sahel Zone - a thorn steppe generally occurring between the 15th and 16th parallels, receiving 200 to 500 mm of rainfall.
3. the Saharan Zone - north of the 16th parallel whose subdesert vegetational communities receive less than 200 mm of rain (Ministere du Plan 1980;26-31).

Soils, formed primarily from the weathering of parent geological material, range from relatively heavy tropical clays in bottomlands to light brown or reddish aeolian sands which are found in dune formations. Regardless of geomorphological origin, these soils vary greatly in: a) their capacities to absorb and retain such moisture as the rainfall regime provides; b) their general level of fertility; c) their potential for sustaining agricultural and pastoral production. Moreover, soils found in the project area are highly susceptible to erosion by wind and rain[2].

Despite recent economic growth, Niger remains one of the world's poorest countries. The per capita gross domestic product is estimated at 61,500 FCFA, or about $US250 (Ministere du Plan 1980;20). Niger's economy achieved an annual growth rate during the 1970s of 7% as a result of the discovery and mining of uranium in desert areas north of Agadez. Certain features of the economy, however, have shown little or no growth within the same period of time. This is especially true of the agricultural and livestock sectors, upon which the majority of the rural populace depends for its livelihood.

The drought of 1969-1973 caused widespread crop failure and devastated the livestock sector of the economy. The government, financed through uranium revenues and substantial amounts of foreign assistance, embarked upon an extensive

Figure 10.1 The Republic of Niger and Tahoua Department

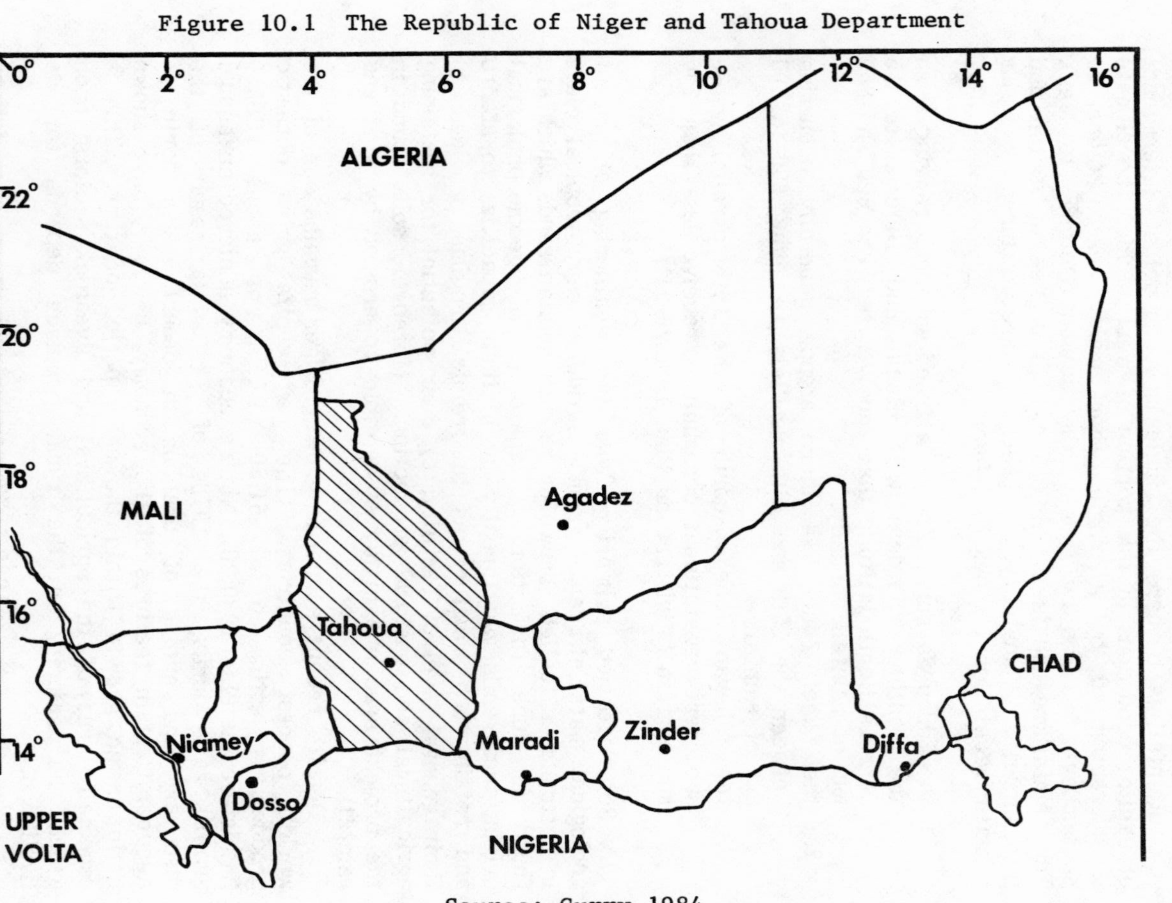

Source: Curry 1984

program of human resource development to rebuild these shattered sectors of Niger's economy. As low density, highly mobile populations inhabiting areas of the country remote from government centers, pastoralists remained largely unaffected by the initial post-drought development efforts of this new program (Curry and Starr 1983;3)[3].

Pastoralists in particular suffered greatly during the drought through substantial losses of their animals, dislocation and resettlement in refugee camps, famine and death. There was a widely held perception of pastoralists as being extremely vulnerable to the vicissitudes of drought. There was, however, little substantive information concerning the ecology of pastoral production systems upon which strategies for increasing pastoralists' capacity to cope with environmental hazards could be based. One of the responses to this knowledge gap was the creation of the Niger Range and Livestock Project in portions of the Agadez, Zinder, Maradi and Tahoua **departments** in the pastoral regions of central Niger.

The Niger Range and Livestock Project

The Niger Range and Livestock Project (NRL) was a collaborative effort by USAID and the Government of Niger's Livestock Service. The purpose of NRL was to provide baseline research on the traditional production and range management systems of pastoralists in Niger (NRL 1976). This information would form the basis of interventions designed to promote the social and economic well-being of pastoralists, while maintaining the rangeland's capacity for sustained utilization by pastoral production systems. In the final stages of the project a series of pilot interventions, based on preliminary findings, were to be instituted. Thus, the project was diagnostic phase of a long-term development effort in Niger's pastoral areas.

The NRL project chose as target populations herders from two pastoral ethnic groups. The Twareg, who form the majority of the population in the NRL project zone, are Berber-speakers (Tamasheq), who traditionally have organized themselves into a series of confederations characterized by a hierarchy of distinct social classes. They migrated to central Niger in successive waves beginning around the 11th century. By contrast, the more recently arrived Wo'daa'be are a relatively unstratified society, where kinship provides the major bond linking different sectors of the population. They lead a highly mobile lifestyle based on small economic units, and socially based forms of mutual aid that serve to minimize risk (Curry and Starr

1983;6-8)[4].

The NRL research program was comprised of several interrelated components. These included: aerial censusing of livestock and compilation of resource inventories for the project zone, vegetative mapping of selected areas, rangeland composition and productivity monitoring, a water point utilization survey, rangeland experiments, veterinary assessment of pastoralists' herds, a livestock marketing study, an inquiry into the legal basis (both traditional and modern) for the formation of herders' associations, and intensive studies of Twareg and Wo'daa'be pastoralists, and of agriculturalists who live within the project zone. While some of these studies were conducted by short-term consultants (i.e. the aerial surveys, veterinary assessment, livestock marketing and legal studies), the majority of the research was undertaken by the project staff, which included both expatriates and their **nigerien** counterparts. In addition, the project sponsored six members of Niger's Livestock Service at universities in the United States to pursue degree training in range management, animal science, and rural sociology[5].

The project research staff was divided into two "teams", one technical, the other socioeconomic. The technical team, consisting of range management and animal production specialists, conducted their research both in the field among pastoralists, and at two government ranches, Ibecetin in the Tahoua **department** and North Dakoro in the Maradi **department**. They were assisted in their research by five Peace Corps volunteers who, along with counterparts and hired herders, monitored the rangeland exclosures and managed the grazing trials at the ranches[6]. The socioeconomic team initiated both extensive and intensive field enquiries, first among Kel-Tamasheq and Wo'daa'be pastoralists throughout the region, and later, among agropastoralists and Hausa-speaking agriculturalists in the vicinity of Shadawanka. All staff members were responsible to the **nigerien** project director for all administrative matters. For the design and implementation of their research, field staff were under the direction of one of two senior consultants in range management and socioeconomics, who visited the project headquarters and field sites several times throughout the year.

On the socioeconomic team were three social and economic anthropologists, each of whom was responsible for a particular aspect of the research. One of the members conducted a survey of water point utilization in the project zone, obtaining data on frequency and seasonality of use of water sources, construction and ownership of wells, seasonal and interannual movements of pastoral herds to exploit water and pasture resources, and access of pastoralists to government services. The other team members conducted field research among Wo'daa'be pastoralists and Hausa farmers. Each of these social scientists brought his or her own theoretical perspective to the collective research effort. This theoretical heterodoxy, which at times sparked lively debate among the research staff, provided the project with a more robust approach to pastoral development than is often the case where there is but a single social scientist on a project[7].

Data Collection Methods

In order to obtain the necessary sociological background information on target groups of pastoralists and to identify constraints to domestic production, the socioeconomic team used both informal and formal data collection methods. In the Shadawanka area, for example, we conducted a series of informal interviews with members of agropastoral hamlets prior to formal censusing to obtain general information on the production system of the group, its history, and local perceptions of constraints to production. Such information was supplemented by periodic discussions with key informants in these groups, a standard ethnographic practice. Examples of formal surveys include the water point survey, which used a single interview questionnaire methodology, and the intensive study of pastoral and agricultural domestic economy, which used a multi-visit strategy.

The latter study was a variant of the "cost route survey" method (Eicher and Baker 1982; 72) to gather data. A random sample of pastoral and agricultural domestic units (whatever their social composition; see Behnke and Kerven 1983 for a discussion of the problems of identifying rural "households" for Farming Systems Research purposes) was interviewed twice weekly concerning labor allocation and budgetary transactions of economically active members. The interview schedules and protocols for the pastoral samples were virtually identical to those used in a similar study conducted by ILCA in the Niger Delta in Mali (Swift; personal communication). Both this system and that used for the agricultural sample were modifications of the cost route survey instruments devised by CRED staff members for their investigations in Upper Volta and Niger. The CRED methodology is described in detail in the report on livestock research in West Africa edited by Shapiro (1979).

The socioeconomic team, under the direction of its senior consultant, utilized the ethnographic knowledge gained from informal survey methods to adapt the CRED cost route methodology to local conditions. Labor data obtained from the twice weekly interviews included such quintessentially pastoral activities such as "going to the bush to search for lost animals" and "making tea" which is an important aspect of Kel-Tamasheq and Wo'daa'be social life. Livestock in pastoral herds were surveyed not only according to age, sex, and species, but also by form of ownership. Economic transactions, particularly exchanges and sales of milk and millet, were recorded according to type and the social relations of the parties involved.

Perhaps the most interesting modification of the CRED system by the project was that of the use of the Tifinagh script to record interviews in Tamasheq and Fulfulde[8]. By using this local alphabet as the medium for data collection, the staff was able to employ herders from the camp under study. They therefore avoided the problem of having to locate and train a cadre of enumerators literate in both French and the pastoral language, who would be regarded in the participating camps as strangers. The data were then numerically coded directly from the Tifinagh and entered on a minicomputer for analysis, thereby eliminating the problem of translation into another language. By permitting herders to serve as enumerators, Tifinagh served, furthermore, to ensure a level of herder participation in the research which is difficult to achieve in surveys of this type.

Introduction to the Tahoua Department

The **department** of Tahoua, where most of the NRL's socioeconomic research was conducted, occupies about 106,677 km², or about 10% of the total surface area of Niger (Ministere du Plan 1980;509). The **department** is subdivided into seven **arrondissements**, the largest of which, Tchin Tabaraden, provided the administrative unit in which the intensive study of domestic economy was conducted. Geologically, there are two major valley systems in Tahoua; the Maggia Valley in the south, the site of most of Niger's cotton production; and part of the Tarka Valley system to the southeast. Relief ranges from 250 m in the southwest to 750 m on the peaks of the Adar-Doutchi Plateau in the south central portion of the **department**. The rest is a series of broad plains, in particular, the Temesna and Tadress plains in the central and eastern parts of the pastoral zone.

Tahoua exhibits nearly the complete range of climatic and vegetational variation found in Niger. The Southern Sahelian zone (350 to 600 m) dominates the southern section as far north as Tahoua and Tassara, which receive between 150 and 350 mm of rain annually, with a climate of the Northern Sahelian type. North of these locales, the climate becomes Saharan, and rainfall averages less than 150 mm. The vegetational regime is roughly congruent with the rainfall and soil patterns. Higher rainfall areas exhibit an arboreal steppe type of plant community, characterized by thorny trees, acacias, some palms and other forbs. Between 250 and 350 mm of rain, arborescent steppes dominated by annual grasses prevail. The more arid regions support gramineous steppe vegetation while the northern portions of Tahoua at times lack vegetation altogether (Curry 1984; 121).

The Population of the Tahoua Department

The sedentary population of 785,640 makes up 79% of the total **department** population, according to the 1977 national census (Ministere du Plan 1980; 510). The mobile pastoral population is estimated at 208,841, or 21% of the total. The sedentary population resides in some 957 villages. These range in size from small hamlets of only a few families, to large villages which contain several thousand inhabitants. The population of Tahoua, like that of Niger, is young and growing. While the official national growth is estimated at 2.7% annually, local growths may in fact be higher. Faulkingham (1977) has estimated the annual rate of growth for a village in Tahoua during the drought to be in excess of 3%. While overall population densities are low when calculated at the **arrondissement** level, local densities of between 52 and 65 persons per km^2 have been reported in the agricultural valleys of the south, where some of the **department**'s most fertile lands exist.

Tahoua is diverse ethnically in both the pastoral and sedentary segments of the population. The pastoral segment is composed primarily of elements of various Kel-Tamasheq (Twareg) confederations, and nomadic Fulfulde speakers (Wo'daa'be; see Dupire 1970). The sedentary population is predominantly Hausa, but is also composed of some Kel-Tamasheq and Igdalen groups, and segments of the Farfaru, Katsinawa and other Ful'be (Fulfulde-speaking) groups[9]. The Hausa who live in the Tahoua trace their ancestry from numerous groups originating in the Adar region **(Adaraawaa)** of Tahoua or from the ancient Hausa kingdom of Gobir **(Goobiiraawaa)**.

Pastoral and Agropastoral Production Systems

In the course of field work, the project social scientists discovered that the agricultural and pastoral groups which they studied utilized a variety of subsistence strategies to gain their livelihood. Despite considerable ethnic and spatial diversity, these subsistence systems can be conveniently classified into several broad categories, according to the relative contribution made by agriculture and animal husbandry to overall production.

Purely pastoral systems are typified by a near total lack of agricultural production. Nomadic pastoralists rely upon the meat and dairy products from domestic herds for subsistence and for exchange to obtain goods which are not produced by the household. Examples of pure pastoralists are to be found among the Kel-Tamasheq and Wo'daa'be groups which exploit the natural resources in the pastoral zone to the north of the village of Shadawanka.

Occasionally, these "pure" pastoralists will engage in a form of agricultural experimentation as a short-term, local response to economic hardship following animal loss due to famine and/or disease. This form of agriculture is characterized by the near total dominance of pastoral production, a low labor input crop, such as millet, a virtual lack of other inputs such as fertilizer, the small size of the holdings, and by relatively low yields.

A more even balance between cultivation and livestock raising can be found among groups of semi-nomadic agriculturalists who use the natural resources of the project area on a permanent or semi-permanent basis. Young men, frequently alone, conduct family herds to rainy season pastures in the northern part of the Tahoua **department**, returning at, or just after, the harvest. Those who remain (i.e. women, old men, and children) reside near the fields in settlements which range in size from adjacent temporary dwellings to permanent villages. They often employ the cultivation technology (seed varieties and hoe types) of their sedentary neighbors.

Semi-nomadic farming systems of this kind can be found among Tamasheq-speaking groups residing near Shadawanka. Some of these people report that, at the end of the first weeding of millet, the older members of the family remain with the very young at the fields to perform the second weeding. Young adults and adolescents follow the herds to the northern pastures to attend the salt cure (**cure sallee**), returning at harvest to tend to the animals in pastures adjacent to the fields. The principle crops grown by Kel-Tamasheq agropastoralists are millet, sorghum and cowpeas.

Sedentary Farming Systems

Sedentary farmers derive the majority of their productive output from agricultural rather than pastoral enterprises to meet their subsistence needs. Herds belonging to sedentary farmers in Shadawanka are composed primarily of sheep and goats (cattle and camels being of less importance) which must rely upon natural forage throughout the year. Hausa sedentary farmers in the Tahoua **department** cultivate a great diversity of crops, both subsistence and cash. These include; numerous local varieties of millet and sorghum, legumes (e.g. cowpeas), cucurbits (calabashes and pumpkins), condiments such as okra and sesame, and cash crops such as cotton, groundnuts and sugar cane. Dry season gardening of cash crops such as tomatoes, onions and fruits both increases market participation and further extends the agricultural calendar to nearly a full-year cycle. As such, it offers a major alternative to dry season wage labor migration for many farmers in southern Niger.

This array of potential crop mixes enables Hausa sedentary farmers to tailor their crop combination on any given field to adjust to local conditions. When rains are late, or a dry spell necessitates replanting, farmers can cope with these environmental problems by seeding with shorter season or more drought tolerant varieties[10].

The primary means of cultivation for Hausa farmers in Tahoua is hand labor. Farmers use bush knife (**adda**) and axe (**gaataari**) to clear fields. The mattock (**kwaasaa**) prepares the field, garden and seedbed, and is the preferred planting implement on rocky soils in areas of the Adar-Doutchi Plateau. More frequently, a narrow bladed planting hoe (**sungumi**) is used for lighter soils. A number of short handled hoes are used for weeding throughout Hausaland; the most common hoe in the Tahoua region is one with a heavy teardrop shaped blade known as **kalmi**. A long handled weeder with a lunate blade (**hawyaa**) is used for superficial weeding of light sandy dune soils in some areas. In the more intensive Hausa farming systems, farmers use animals to power clearing, planting and cultivation operations. This component of the farming system is, however, totally lacking in Shadawanka.

Sedentary Production Systems

Sedentary farmers of Tahoua use this hand technology to exploit a variety of landforms. They include dune **(tudu)**, plateau **(dabagi)**, areas of undulating hillocks **(tuduni)**, and bottomlands **(fadama)** on valley floors. In some areas, **fadamas** can support continuous cropping of cereals and cash crops in the rainy season, and vegetable gardens irrigated by shallow hand-dug wells and water lifts in the dry season.

The agricultural cycle begins with the clearing of fields in April or May, a month or two prior to the start of the rains. Seeding occurs after the rains have begun in earnest. Teams of two plant the grain; one person makes holes in the ground two paces apart with a hoe, the other drops in a few seeds, and covers them with earth. Households prefer to rely upon family labor to sow millet and sorghum, since the irregular pattern of rainfall in the region often necessitates replanting.

The first weeding occurs a couple of weeks later, after the plants have reached several centimeters in height. At about this time, the millet and sorghum may be intercropped with cowpeas, and later pumpkins. Weeding continues throughout the rainy season. Farmers average two to three weedings per field, depending upon conditions. It is during weeding that hiring and exchange of labor most often occur. The weeding is sometimes organized into work parties **(gayyaa)** in which the participants are provided with meals and small gifts[11].

The end of the rains in September signals the end of the weeding and beginning of the harvest. As the grain heads ripen, children spend much time in fields guarding the harvest from weaver birds. Pannicles of millet are harvested with a small knife. The heads are allowed to dry in the fields, then bundled into headloads for transport to the households' granaries. The grain harvest may last from September to November; cowpeas and pumpkins may extend the work until well into December.

At the conclusion of harvest, farmers begin their dry season round of activities. Since harvests in this region of Niger are seldom sufficient to sustain households for the entire year, off-farm income generating activities predominate. Some farmers in possession of a hereditary craft such as blacksmithing or tanning ply their trades. If in possession of bottomland, they may begin preparations for an irrigated garden of cash crops such as onions, tomatoes, or fruit trees. Many Hausa engage in commerce of goods and livestock on a local or regional basis[12]. Those without such options are forced to migrate to cities such as Lagos on the coast in search of wage employment.

Access to farmland is established by clearing either new or existing farmland of bush, and by keeping the land in more or less continuous cultivation. In many areas of Tahoua, high population densities virtually assure that little "new" bush is available. Transference of use rights can occur through inheritance (**gaadoo**), borrowing or pledging (**aro**), gifts (**kyautaa**), and entrustment (**amaana**). Leasing of land (**haya**) and sales of use rights (**sayon goona**) are increasing throughout Hausaland as a result of increased monetization of village economies (Raynaut 1976, 1977).

The Village of Shadawanka

The village of Shadawanka (Figure 10.2) is located north of the Adar-Doutchi Plateau, and at the foot of the pastoral areas of the Azawak region in the Tahoua **department.** It is, in reality, two villages, separated by the market on an island in a seasonal water course, each with its village head (**hakimi**). The older, western portion of Shadawanka is composed of Hausa from the Adar region near the town of Keita, who informants say settled there shortly after the establishment of the market at its present location, less than forty years ago. Residents of this segment of the village still consider themselves as subject to the national and traditional authorities in Keita, to whom they paid the head tax prior to its abolition in the mid-1970s.

The eastern half of the village is, however, subject to the jurisdiction of the authorities of the administrative post of Abalak, some 65 km to the northeast. The majority of the Hausa families arrived from the Adar or the Gobir less than forty years ago. In addition, there are a few Ful'be and Kel-Tamasheq households. It is this eastern half of the village where the NRL socioeconomic inquiries were conducted. All subsequent references to Shadawanka will be to this portion of the village.

The population of Shadawanka was, at the beginning of the NRL study in July 1981, 429 persons; 215 males and 214 females. The principal unit of production and consumption of the village is the household (**gida**). The 66 households censused exhibited considerable variation in size and composition, and included both simple (**iiyaali**) and compound (**gandu**) types of organization. **Iiyaali** households, both monogamous and polygynous, were the most common.

The farming system of Shadawanka is best considered as an

Figure 10.2
The Village of Shadawanka and the Research Area

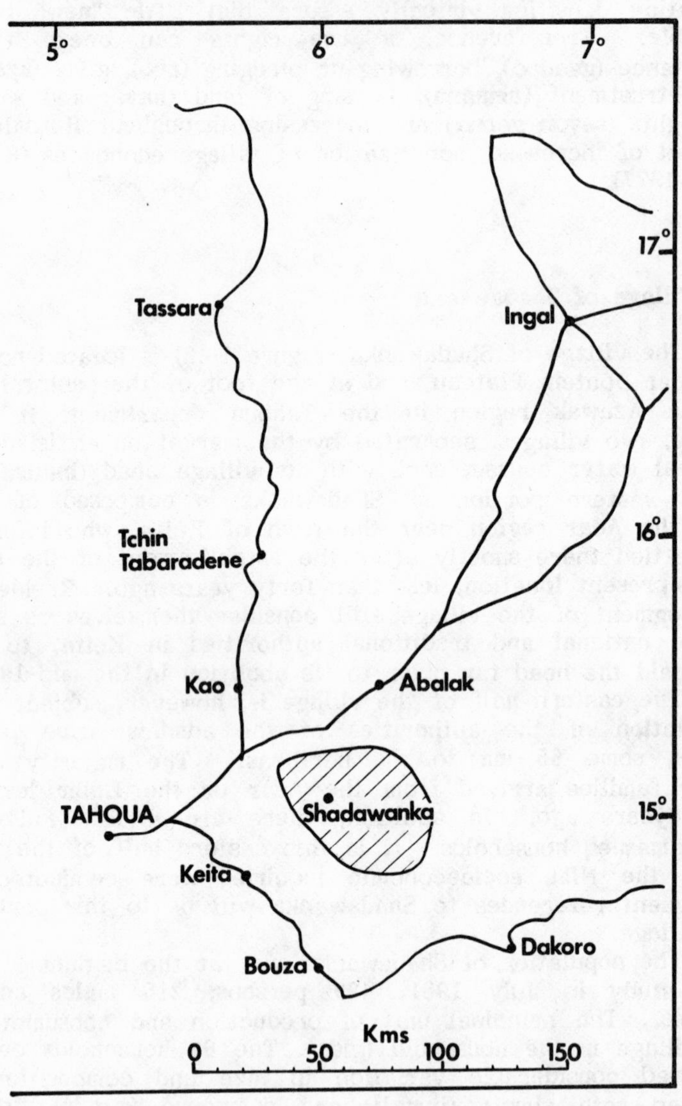

Table 10.1 Characteristics of Sedentary Agricultural Villages and Regions, from North to South, Niger and Nigeria.

Zone Village/State	# of households (HH)	Hectares cultivated /HH	Fields / HH	Average Field Size	Persons / Ha.	Ha/ Person	Active /Ha	Has. Active
Transitional								
Shadawanka [1]	21	3.2	1.95	1.91	6.64	0.64	3.24	1.10
Agropastoral								
Dan Koullou [2]	34	6.6	NA	NA	NA	NA	5.00	1.40
"Z" [2]	48	16.9	2.40	7.20	6.50	NA	0.38	4.40
Sherkin Hausa	7	10.0	NA	NA	9.00	NA	5.30	2.00
Agricultural								
Magaria [1]	23	3.6	3.91	0.91	6.96	0.60	4.61	1.11
Pure Agricultural								
Magami [2]	8	5.2	NA	NA	7.60	NA	4.40	1.80
"Y"[2]	44	7.4	1.00	7.40	6.90	NA	4.00	1.90
Nigeria								
Kaita [3]	473	3.9	2.40	1.60	5.90	0.66	3.35	1.16
Sokoto [4]	99	3.9	5.80	0.67	5.60	0.70	1.50	1.05
Zaria [4]	124	3.9	6.20	0.63	8.60	0.50	2.20	1.10
Bauchi	116	3.9	6.80	0.57	6.00	0.70	1.60	2.12

1. From fieldwork (1981).
2. From Sutter (1979).
3. Derived from Watts (1979;426,431,432).
4. Derived from Norman, Pryor and Gibbs (1979), various tables.

abbreviated, and arid, variant of the general Hausa farming pattern described previously. The agricultural landscape surrounding the village consists of semi-stable sand dunes whose soils can sustain millet cultivation, and low-lying runoff areas which become inundated during the rainy season and can support sorghum production. Since the area around Shadawanka receives between 300 and 400 mm of rainfall annually, climatic conditions place the village and its environs at the extreme northern limit of dryland cereal cultivation in Hausaland.

Table 10.1 summarizes some of the principal characteristics of farming units in the Shadawanka study sample and compares them to other Hausa farming situations in northern Nigeria and southern Niger. Nearly all households in the village possess some farmland[13]. The mean plot size for the study sample, 1.91 ha, is considerably higher than that reported for northern Nigeria by Norman, Pryor and Gibbs (1979;26), but lower than values reported for other agropastoral locations in Niger. The total hectares cultivated, however, is the lowest of any of the villages reported in the table. The average number of fields per household is approximately two. Shadawanka's farmers cultivate five different crop mixtures of millet, sorghum and cowpeas. These are considerably fewer combinations than those reported for Zaria by Norman, Simmons and Hays (1982;52). As Table 10.2 indicates, sole cropping of millet is the preferred production pattern. This is understandable since the majority of the fields are located on sandy dune soils which are suitable for millet.

Table 10.2
Distribution of Fields by Crop Type; Shadawanka, 1981.

Crop Mix	Fields	Hectares	Mean Plot Size
Millet, sole crop	14	20.57	1.47
Sorghum, sole crop	4	8.13	2.03
Millet-Sorghum	10	23.74	2.37
Millet-Sorghum-Cowpeas	3	8.80	2.93
Millet-Cowpeas	7	9.97	1.40
Fallow	2	4.25	2.13
TOTAL	40	75.46	2.06

Source: Curry 1984.

Farmers in Shadawanka employ the hand cultivation technology common to other Hausa farmers; planting hoe (**sungumi**), weeding hoe (**kalmi**), mattock (**kwaasaa**), and bush knife. Conspicuously absent from the farming system is the use of animal traction for land preparation, planting and cultivating. Climatic conditions, specifically a short growing season (approximately three months) and the constant threat of crop failure due to unpredictable rains, place severe limitations upon this hand cultivation of dryland cereals.

Consequently, farmers eschew the use of any inputs of fertilizer, organic or chemical. When asked, informants stated that they felt that such a practice would burn the millet and sorghum, given the generally "hot" soils they farmed[14]. The only available source of soil fertility maintenance is manure from livestock belonging to sedentary and pastoral herds which are permitted to graze on fields during the dry season. Such a practice is common throughout West Africa and is at times cited as an example of the symbiosis of crop/livestock interactions in these farming systems. Using a method devised in collaboration with the range management staff, we estimated the amount of manure available from this practice on fields of the sample households. The thirty-seven fields measured had an average of 753 kg per ha. Such a low figure suggests that, in the Shadawanka case, the crop/livestock interaction in the farming system provided little benefit to farmers **vis a vis** soil fertility maintenance[15].

It is not surprising, then, that yields from this farming system are quite low. For the 1981 harvest, we estimated that average yield for millet and sorghum was about 287 kg/ha, with a range of 56 to 1685 kg/ha (Curry 1984;236). The average is substantially lower than the 366 kg/ha for millet and 786 kg/ha for sorghum reported for northern Nigeria (Norman, Simmons and Hays 1982; 236). For this meager harvest, households in Shadawanka devote considerable time and effort to farming. Table 10.3 compares the total annual hours spent on the family farm and the source of this labor by the Shadawanka sample with that for areas of Northern Nigeria. It is apparent from the yield and labor data that farmers spend nearly the same amount of time in farming for less than half the reward as do their counterparts to the south.

The poor performance of farming in Shadawanka is even more apparent when one considers its capacity to feed those who participate in it. Table 10.4 provides an estimate of yields for the sample households in the years 1980 to 1982. We calculated average daily household consumption for the sample based on data gathered during both the rainy and dry seasons in 1981. From these data, we determined that households could support themselves with these harvests between 114 and 140 days, less than half of the year. Consequently, the people of Shadawanka must rely heavily on purchases of staple grains to to sustain themselves throughout the rest of the year.

Table 10.3
Composition of the Agricultural Labor Force.
Shadawanka and Northern Nigeria

	Northern Nigeria			Shadawanka	
	Sokoto	Zaria	Bauchi	1981	1982
Annual hours on Family Farm	1566.0	1800.0	1317.0	1219	1348
Source of Farmwork % of total hours					
Family					
Adult male	72.4	72.2	73.3	44	40
Adult female	1.1	0.3	1.6	17	15
Children	8.5	8.9	9.4	37	35
Salaried and Invited Labor	18.0	18.6	16.0	3	8

Source: Curry 1984.

In order to meet their subsistence needs, villagers obtain cash through a variety of occupational strategies. These include sales of traditional crafts, petty commerce, participation in the livestock trade as buyers, sellers and middlemen (**dillaalai**), and wage employment within the local and regional economies. Of particular importance for the livestock project is the place of Shadawanka's livestock trade in the domestic economies of both farmers and herders in the region. The weekly village market is one of the major centers for livestock trade in the area and, along with those of Kao and Abalak, forms an important commercial link in the long livestock marketing circuit between pastoral producers in the Azawak region and consumer markets which stretch as far as Lagos to the south (Makinon and Arizo-Nino 1982). Several village residents participate in the livestock market as middlemen, who mediate livestock transactions. Others buy livestock locally at the beginning of the dry season and drive them to the border with Nigeria for sale. Even the major traditional craft practiced in the village, tanning, is an integral part of the livestock sector of the village economy.

The importance of trade for the people of Shadawanka can be traced to the origins of the village itself, which according to local informants was founded by a Hausa trader engaged in commerce with the pastoralists of the region. Trade, therefore, rather than local production, has been the **raison d'etre** for the existence of sedentary agriculture in Shadawanka, and has both shaped and been shaped by local constraints to production, and the socioeconomic relations between farmers and herders in the region.

Table 10.4
Consumption of Grain by Sample Households
Shadawanka

Year	1980	1981	1982
Kg of grain harvested per household.	676	817	690
Average Daily consumption (kg)	--	5.36	--
Number of Days of grain consumption from harvest	118	140	114

Source: Curry 1984.

From these results, several conclusions were drawn concerning the nature of the relationships between farmers and herders in the project zone. Agriculture appears to be an underproductive subsistence strategy in the study area that competes directly with pastoralism for the region's land resources. In the NRL project area, farming persists well to the north of the 15th parallel, the line designated as the legal limit for agriculture. Farming, it was concluded, posed a threat to the continued persistence of pastoralism in the project area through its conversion of productive grazing land to unproductive farmland. Consequently, the project supported efforts by the government to discourage encroachment of pastoral areas by agriculturalists.

Conclusions

The Niger Range and Livestock Project utilized a systems approach similar to that found in FSR/E efforts in its attempt to understand the pastoral ecology of central Niger. Although it conducted the greatest proportion of its technical research on-ranch, the project staff devoted considerable time and effort to field research involving the participation of herders and farmers in the project area. The socioeconomic team composed primarily of social and economic anthropologists used a variant of the cost route survey method employed in adaptive research by Norman and others to examine pastoral and agricultural production in the context of domestic economy. Thus, the project owed much of the design of its research program to previous work in farming systems.

This data collection program proved successful in several ways. The surveys generated a substantial amount of both qualitative and quantitative data on pastoralists and agriculturalists in central Niger. The formal structure of the single-visit water point survey and the multi-visit study of the domestic economy offered staff anthropologists an **entree** into numerous pastoral camps and farm villages where they collected the supporting qualitative data necessary for the interpretation of the quantitative results of the surveys. The rapport established by the project with target groups of pastoralists through such long term contact proved invaluable in the implementation of pilot interventions with experimental herders' associations at the conclusion of the project. Through their participation in the research, members of the Nigerien staff received training in socioeconomic research methods.

However, the cost route survey method itself proved to be a serious problem in the NRL's experiment in pastoral systems research. The survey was costly in terms of both resources and time devoted to it by project staff. Numerous person months were spent collecting, coding, entering and verifying the rather enormous volume of data generated. Limitations of computer hardware and software, including capacity and availability, hindered completion of analysis by the conclusion of the project. To date, much of the data remain to be analyzed. Eicher and Baker's observation that "processing of survey data ..[poses] .. a major problem for researchers throughout Africa" (1982;77) certainly characterizes the situation which confronted NRL at the conclusion of the formal surveys[16].

The "data crunch" problem posed particular difficulties for the designers of the successor projects, who arrived within three months after cessation of field activities in October 1982. With the entry and analysis of the survey data far from complete, the robust quantitative results anticipated from the NRL were unavailable for planning the new project. Final reports by some of the research staff were at that time in draft form.

Faced with this situation, the team was required to rely upon the qualitative data gathered by NRL staff members during field work in its design efforts. These qualitative data included a detailed study of the traditional herding practices of the Wo'daa'be by Maliki (1982), based on his long term experiences with these herders, and numerous consultations with the social anthropologist studying the Wo'daa'be. It is interesting to note that, in this situation, the ethnographic expertise displayed by the staff anthropologists, particularly the Wo'daa'be specialist, gave them sufficient credibility to allow them to participate in the design process, thereby affecting policy decisions for the subsequent project.

Anthropologists contributed to development planning of pastoral areas in several ways. The project model for organization of pastoral associations, contained in a position paper written by the pastoral specialists, and based on staff field experiences, was subsequently incorporated into government guidelines for the development of pastoral areas. One of the economic interventions prominent in the design of the successor project is a revolving credit fund for association members. This fund is modeled after the traditional system of entrustment of cattle used by the Wo'daa'be[17]. During the pilot intervention phase at the end of NRL, some associations were permitted to maintain their records in the Tifinagh, rather than in Roman script. This is, I feel, an outcome of project experience with the use of Tifinagh to gather economic data during field work.

In the NRL case, therefore, it was more through informal consultation, rather than by formal analysis of the project's data, that anthropologists were able to influence policy decisions. Because of their skills in collecting and interpreting qualitative data, and of their familiarity with the "local situation", the NRL anthropologists became, for the policy makers, credible informants and valuable sources of information upon which policy recommendations could be based.

The experiences of the Niger Range and Livestock Project serve as a reminder that anthropology has much to contribute to a systems oriented research program. Through their training, anthropologists approach a multidisciplinary research situation with a keen awareness of the interrelationships between the social and the technical, and with a set of analytical and methodological tools which enable them to describe, albeit qualitatively, and to understand the essential features of the system under investigation. It is, at times, this qualitative information collected by anthropologists in the course of their field work that provides the basis, and sometimes the only basis, upon which sound development policy can be fashioned.

Notes

1. The field research which I conducted as a member of the Niger Range and Livestock Project, 683-0202, was carried out under USAID contract No. 683-A-10003 and through research permission granted by the **Institute des Recherches en Sciences Humaines, Universite de Niamey.** The author gratefully acknowledges their support. All statements in this paper are the sole responsibility of the author.
2. Soil erosion losses in the Maggia Valley just to the south of Shadawanka, for example, have been estimated to vary from 3.5 to 18.5 tons per ha annually (Curry 1984;98).
3. Development efforts in the pastoral areas during this period of infrastructural improvement consisted primarily of drilling wells, the inoculation of livestock against rinderpest, and, to a lesser extent, the establishment of "nomad schools" **(ecoles nomades)** and dispensaries at government outposts.
4. For definitive ethnographic treatment of Kel-Tamasheq societies in Niger, see Bernus (1981). Dupire (1970,1972) presents an excellent overview of Ful'be groups, including the Wo'daa'be. A history of the Wo'daa'be in Niger has been written by Maliki (1981).
5. Two of the degree trainees returned to the project during the last year to participate in field activities. Their experience was indispensible to the process of transition from NRL to the successor project, the Niger Integrated Livestock Production Project (NILPP). In February of 1984, the new project director appointed them coordinators of the technical research and pastoral coordination sections of NILPP.
6. Two of these volunteers returned, upon completion of Peace Corps service, as technical staff assistants for range management and animal production for the final 15 months of the project.
7. The remainder of the team consisted of: a pastoral economist (Senior Consultant), a physician, and an ex-Peace Corps volunteer and a Catholic priest whose familiarity with the language and culture of the Twareg and Wo'daa'be provided additional ethnographic expertise.
8. Tifinagh is used traditionally by Tamasheq-speaking artisans to sign their work and by members of nearly all classes to send love letters. The water point survey data revealed a high degree of literacy among both Twareg and, to a lesser extent, Wo'daa'be in Tifinagh. It was consequently

adopted as the written medium for the pastoral domestic survey.

9. The term "sedentary" here implies the use of a permanent or semi-permanent dwelling. In fact, many of the Kel-Tamasheq and Ful'be agropastoralists enumerated under the sedentary population category in government are seasonally quite mobile.

10. For an illuminating discussion of the complexities of Hausa farming systems in Kaduna State, Northern Nigeria, see Watts (1979).

11. In many areas of Hausaland, the practice of **gayaa** seems to be in the process of being either transformed through monetization, or replaced altogether by wage labor as the preferred form of extra-household recruitment of agricultural labor (cf. Raynaut 1976, 1977; Sutter 1979, 1982; Watts 1979).

12. Long-distance trade has long been an integral feature of Hausa society. The cattle-kola circuit, for example, is well documented (cf. Baier 1982; Curry 1984; Lovejoy and Baier 1975; Makinon and Arizo-Nino 1982).

13. There were two exceptions. One family of Ful'be earned their living as herders for the village, and did not farm. The other household consisted of a single, elderly woman, who subsisted upon gifts of food and other commodities from neighbors.

14. The "hot/cold" distinction of soils in Shadawanka and elsewhere in Tahoua appears from informants' statements to refer to the soil's temperature, and its capacity to absorb and retain moisture, virtually regardless of soil type and structure. This moisture utilization feature of local soil taxonomy is, I feel, an important diagnostic tool in farming systems of the region.

15. For an analysis of farmer/herder interactions in another part of the Sahel, see Delgado's chapter in Shapiro (1979).

16. Data entry problems were legion. Data were collected and coded in Niger by project staff, both in the field and in Niamey. Data were entered both in Niamey and in the US. Computer "crashes" in Niamey and irregular shipments of entered data to NRL from the US caused considerable delay in completion of the data set. The inability of NRL to complete data entry and clean up required the contracting out for this service at project's end. This work was further hampered by delays of shipment of the data forms from Niger, and by the project's losing the original data sheets for the Twareg samples.

17. For a detailed description of this and other aspects of Wo'daa'be production, see Maliki (1982).

Bibliography

Baier, Stephen. 1980.
An Economic History of Central Niger. Oxford: Oxford University Press.

Behnke, Roy and Carol Kerven. 1983.
"FSR and the Attempt to Understand the Goals and Motivations of Farmers". In **Culture and Agriculture,** Issue 19, Spring 1983;9-16.

Bernus, Edmond. 1981.
Touaregs Nigeriens: Unite Culturelle et Diversite Regionale d'un Peuple Pasteur. Paris: Memoires ORSTOM #94.

Cohen, Abner. 1965.
"The Social Organization of Credit in a West African Cattle Market". **Africa** 35(1); 18-36.

——. 1966. "The Politics of the Kola Trade." **Africa** 36(1); 18-36.

——. 1969. **Custom and Politics in Urban Africa: A Study of Hausa Migrants in Yoruba Towns.** London. Routledge and Kegan Paul.

——. 1971. "Cultural Strategies in the Organization of Trading Diasporas". In **The Development of Indigenous Trade and Markets in West Africa.** C. Meillassoux (ed.). London: IAI, Oxford University Press; 267-281.

Curry, John James, Jr. 1984.
Local Production, Regional Commerce, and Social Differentiation in a Hausa Village in Niger. Ph.D. dissertation, Department of Anthropology, University of Massachusetts.

Curry, John James, Jr., and M. Starr. 1983.
The Niger Range and Livestock Project: A Case Study of Women and Development. Washington DC: USAID.

Dupire, Marguerite. 1970.
Organisation Sociale des Peul. Paris: Librairie Plon.

——. 1972. **Les Facteurs Humains de L'Economie Pastorale.** Niamey: CNRSH, Etudes Nigeriennes #6.

Eicher, Carl K. and D.C. Baker. 1982.
Research in Sub-Saharan Africa: A Critical Survey. Michigan State University International Development Paper #6. East Lansing: Department of Agricultural Economics, Michigan State University.

Faulkingham, Ralph H. 1977.
"Ecologic Constraints and Subsistence Strategies: The Impact of Drought in a Hausa Village, a Case Study from Niger." In **Drought in Africa.** D. Dalby, R.J. Harrison-Church and F.Bazzaz (eds.). London: IAI Oxford University Press.

Goldschmidt, Walter R. 1975.
"A National Livestock Bank; An Institutional Device for Rationalizing the Economy of Tribal Pastoralists". In **International Development Review,** 2; 2-6.

ILCA (International Livestock Centre for Africa). 1984.
Annual Report. Addis Ababa: ILCA.

Lovejoy, Paul E. 1974.
"Interregional Monetary Flows in the Pre-Colonial Trade of Nigeria." **Journal of African History** XIV(4); 633-651.

Lovejoy, Paul E. and S. Baier. 1975.
"The Desert-Side Economy of the Central Sudan." **International Journal of African Historical Studies** 8; 551-581.

Makinon, Marty and E. Arizo-Nino. 1982.
The Market for Livestock for the Central Niger Zone. Ann Arbor, Michigan: Center for Research on Economic Development, University of Michigan.

Maliki, Angelo. 1981.
Historique du Wodaabe. Niamey: USAID.

——. 1982. **Nganyaka.** Niamey: USAID.

McDowell, R.E. and P.E. Hildebrand. 1980.
Integrated Crop and Animal Production: Making the Most of Resources Available to Small Farms in Developing Countries. New York: Rockefeller Foundation.

Ministere du Plan. 1980.
Plan Quinquennal de Developpement Economique et Social 1979-1983. Niamey: L'Imprimerie Nationale du Niger.

Norman, David W., I.H. Pryor, and C.J.N. Gibbs. 1979.
Technical Change and the Small Farmer in Hausaland, Northern Nigeria. African Rural

Economy Paper #21, East Lansing: Department of Agricultural Economics, Michigan State University.

Norman, David W., E.B. Simmons, and H. Hays. 1982. **Farming Systems in the Nigerian Savannah: Research Strategies for Development.** Boulder Colorado: Westview Press.

NRL. 1976. **Project Paper.** Niamey: USAID.

Raynaut, Claude. 1976. "Transformation du Systeme de Production et Inegalite Economique: Le Cas d'un Village Haoussa (Niger)." **Canadian Journal of African Studies** X(2); 279-306.

——. 1977. "Circulation monetaire et evolution des structures socioeconomiques chez les Haoussas du Niger." **Africa** 47(2); 160-171.

Shapiro, Kenneth H. (ed). 1979. **Livestock Production and Marketing in the Entente States of West Africa: Summary Report.** Ann Arbor, Michigan: Center for Research on Economic Development, University of Michigan.

Sutter, John W. 1979. "Social Analysis of the Nigerien Rural Producer." In **Niger Agricultural Sector Assessment.** Vols. I and II. W.J. Enger and M. Smale (eds.). Niamey: USAID.

——. 1982. "Commercial Strategies, Drought, and Monetary Pressure; WoDaaBe Nomads of the Tahout Arrondissement, Niger." **Nomadic Peoples** 11; 26-60.

Watts, Michael J. 1979. **A Silent Revolution; The Nature of Famine and the Changing Character of Food Production in Nigerian Hausaland.** Ph.D. Dissertation Department of Geography. University of Michigan.

PART V

CASE STUDIES IN AGROFORESTRY

11

Indigenous Technology and Farming Systems Research: Agroforestry in the Indian Desert[1]

Barry H. Michie

The appreciation of indigenous technology is an underlying tenet of Farming Systems Research. Traditional agricultural production systems in many cases are adaptations to local social and ecological conditions which have proven to be productive and sustainable over long periods of time. Agricultural science has investigated relatively few farming systems, and as a result the observation and study of undocumented systems may be a necessary prerequisite to the improvement of local systems. The application of anthropological techniques can be a valuable component of this initial observation, particularly for eliciting the rationale of existing systems. The application of a holistic perspective can be an important tool for bringing together technical and human elements for the generation of new technology or the improvement of existing systems.

This case study describes a Farming Systems project directed by the author in India with the faculty of Mohan Lal Sukhadia Agricultural University[2], the agricultural school for the State of Rajasthan. The project dealt with an arid zone where grain is produced in association nitrogen fixing trees. The multiple benefits of this association make it unlikely that farmers will change this production system unless major climatic or infrastructural changes occur. The probable persistence of this pattern has major implications for agricultural research strategies.

Location Specificity in Farming Systems Research: The Case of India

Farming Systems Research is fundamentally a location

specific methodology. Soils, climate, topography, cropping patterns, floral and faunal components form the bio-physical elements of an agricultural system, but even within so-called homogeneous cropping zones (GOI 1976; Morrison 1979) the diversity is great due to the variety of associations which can be formed with that set of elements. Furthermore the diversity in the bio-physical dimension is mirrored in the socioeconomic arrangements through which the bio-physical ones are organized and acted upon (Chambers and Harriss 1977; Mencher 1966; Nair 1979). The obvious implication is that what might work in one environment or system will not work necessarily in another.

In India there is explicit recognition of this diversity and attention is turning to location specific research. The fifteen volume **1976 National Commission on Agriculture Report** published by the Ministry of Agriculture and Irrigation (GOI 1976) identifies some 11 rainfall, 52 cropping and 29 livestock zones for Rajasthan, the location of the case study (GOI 1977).

Rajasthan is in the northwest part of India (see Figure 11.1), covers some 132,000 square miles, and is India's second largest state. To the west is Pakistan with the international border running through the Thar or Great Indian Desert. The major topographic feature affecting rainfall and cropping patterns is the Aravalli Mountains that enter the south central part of the state and continue in a northeasterly direction through the neighboring state of Haryana and end near Delhi. The bulk of the state lies west of these mountains and is marked by semi-arid to arid conditions. The climate is dominated by the monsoon with rainfall concentrated in the hot summer months of June through September. This is the main cropping season, called **kharif**. A second cropping season, called **rabi**, occurs during the dry cool winter months and is dependent on irrigation, light winter showers, and/or moisture retention of soils. The southeastern part of the the state receives the highest rainfall (up to 40 inches), with decreasing amounts as one moves to the west-northwest. Lowest rainfall occurs in the westernmost tracts where in places it averages less than 4 inches per year. Moving from the south and southeast to the northwest, rainfed **kharif** cropping shades from maize into sorghum and then into pearl millet. The decreasing moisture requirements of these crops reflect the diminishing average rainfall. Add to this the **rabi** crops of wheat and barley and local variations due to irrigation (particularly in the canal areas of the northwest) and soils, and the agro-climatic map or Rajasthan is quite diverse (GOR 1983; ICAR 1980).

Environmental specificity appears to be a major concern in Indian agricultural research and development planning. Not only is this reflected in the various national and state reports of the

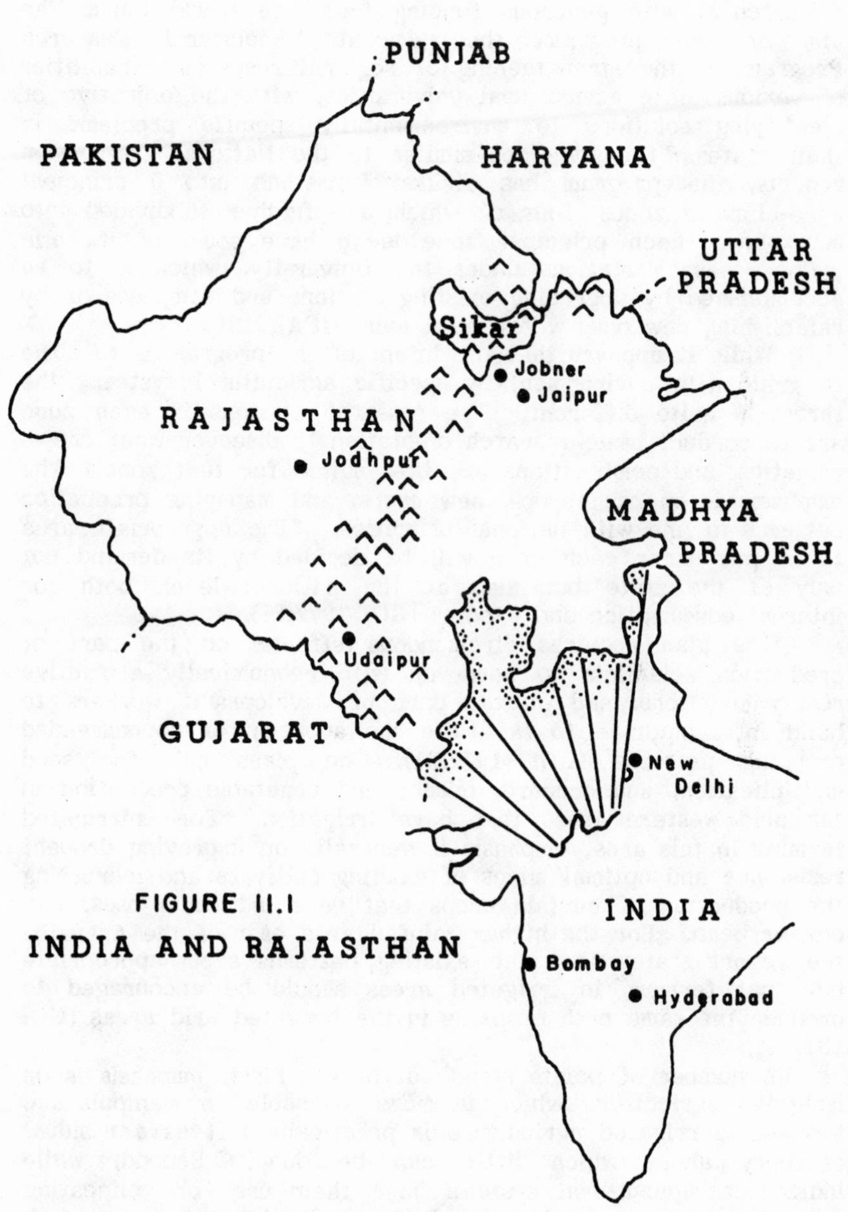

FIGURE 11.1
INDIA AND RAJASTHAN

National Commission on Agriculture but also in a major research program initiated by the Indian Council of Agricultural Research[3] with generous funding from the World Bank. The aim of this program, the national Agricultural Research Program, is the strengthening of regional research capabilities of various state agricultural universities, with the objective of identifying solutions to environmentally specific problems in their states. On guidelines similar to the National Commission reports, this program has divided Rajasthan into 5 principal agro-climatic zones, four of which are further subdivided into sub-zones. Each principal zone is to have some of its own major research stations under the University, which is to be accomplished by upgrading existing stations and campuses or by establishing new ones where none exist (ICAR 1980).

While it appears that the intent of the program is to come to grips with environmentally specific agricultural systems, the thrust is quite different. The research stations in each zone are to conduct basic research on-station to discover what crops, varieties, and combinations are best suited for that zone. The emphasis is on cash crops, new crops, and managing production patterns in line with national priorities. "The appropriate area to be put under each crop will be decided by its demand not only at the state but also at the national level, both for internal consumption and export" (GOI 1977;21).

The plan requires "tremendous efforts" on the part of production scientists to come up with economically attractive recommendations, and on the part of development workers to build infrastructure to facilitate the adoption of recommended cropping patterns (GOI 1977;21). The plans call for seed multiplication, and orchard, oilseed and vegetable production in the arid western areas that have irrigation. For unirrigated farming in this area, emphasis is generally on improving drought resistance and optimal mixes of existing cultivars and increasing the production of certain crops that have industrial uses, e.g. cluster bean. For the higher rainfall area east of the Aravallis the report states that the existing patterns seem appropriate but that farmers in irrigated areas should be encouraged to produce the same cash crops as in the irrigated arid areas (GOI 1977;22).

A number of points stand out here. First, emphasis is on irrigated agriculture which is more amenable to manipulation. Dryland unirrigated agriculture is practically left as a residual category about which little can be done. Second, while indigenous agricultural systems have their use for delineating agro-climatic zones and major features for laboratory research plot experimentation, these systems are seen as having nothing else to offer. This is a recurring theme, not particularly unique to India but implicit in much development and applied research policy. Traditional systems are by definition backward, unproductive and inefficient. Since they are to be changed and replaced through the application of science, there is no need to interact or understand them except at the extension stage. Recommendations and packages of practices are developed on research stations and taken to farmers who, rational creatures that they are, will accept the self-evident advantages of "scientific" farming.

The basic argument of FSR is to the contrary. There is much to be learned about how resources are combined in traditional systems, in this case dryland rainfed ones that have practical value for research directed to increasing production under such conditions. The associations and relationships of the bio-physical elements that have been brought into being by farmers often hold keys to solutions that cannot be discovered through the **a priori** assumptions and methodologies of conventional on-station disciplinary research.

This is particularly imperative in a state such as Rajasthan where most agriculture is practiced in a fragile environment, where interdependencies are critical and immediate, and where population pressure will lead to attempts to intensify production regardless of what research and development programs are doing. While moisture is the major regulating factor for crop production in a desert environment, rainfed systems should not be sold short for their potential.

Rajasthan's agriculture is predominantly rainfed. Estimates of gross irrigated area as a percentage of gross cropped area are 16 to 18 percent depending on the source (GOI 1977; ICAR 1980); most is concentrated east of the Aravalli Mountains, in their southern range, or in the canal irrigated areas of the northwest. As for the arid to semi-arid zones, particularly west of Aravallis, the main physical and climatological features of rainfed agriculture comprise a rather formidable list of difficulties; sandy to loamy soils with low moisture retention and low organic matter content, high growing season temperatures, dessicating winds, and low and erratic rainfall. Rainfed agriculture is subject to frequent failure and low yields due to these factors.

Combined with these bio-physical features, the human element also provides a set of problems unique to the Rajasthan desert. Of all desert zones, Rajasthan has one of the highest population densities in the world. The 61% of the state's area that lies west of the Aravallis contains 13.39 million people, or 39% of the total population, for an average density of 167 per square mile (GOR 1983). Population increase in western Rajasthan is higher than the rest of the state and exceeds the all-India average as well. Perversely, there is an inverse relationship between population growth and amount of annual rainfall (Mann 1981;482-483). This population growth exerts increasing pressure on natural resources for food and fuel for humans, and fodder for an animal population increasing at a faster rate than the human population. Intensification of production is the inevitable result, through bringing more marginal lands into production, the use of shorter fallows, and the extraction of resources from common lands that leads to deforestation and pasture degradation. The problem, that is also a crisis, is how to meet these demands without further degrading the environment. An understanding of these systems is crucial.

As is often the case, agricultural systems are not seen usually as systems at all, but only as fractured pieces of a whole (Morrison 1979; 633). This is the case for the rainfed systems of western Rajasthan, although perhaps less so for irrigated areas across the state that have received much more attention. For example Gupta and Prakash's edited volume, **Environmental Analysis of the Thar Desert** (1975), contains exhaustive and very good descriptive accounts of various features of the area from a wide range of disciplines. However there is no analysis linking them together, let alone how natural resources are used, combined and exploited in actual production systems.

Similarly H.S. Mann, former Director General of the Central Arid Zone Research Institute in Jodhpur, discusses the research program of the institute and the environmental conditions in which the program is working. He describes the various problematic features outlined above, agronomic experiments the institute has conducted, and a few recommendations that have developed for crop mixes, moisture conservation, and cultivation practices. However, the emphasis is on field crops (only part of the system), on-station research and no discussion of what is going on in the region's agricultural systems other than an enumeration of human and livestock populations, crops grown and their low yields, and the dangers of resource degradation. Significantly he does call for interdisciplinary efforts but these are for limited surveys of physical features and interpretation of remote sensing data (Mann 1981). While on-station research is desirable and necessary, there must be an understanding of how people "out there" cope and adapt to such conditions and changes with their production strategies. This also underscores a point that must be made that research stations located in specific zones are no guarantee of an increase in the knowledge and understanding of location specific agricultural systems.

The Farming Systems Case Study

Recently the author was involved in a FSR project with Mohal Lal Sukhadia Agricultural University coordinated through the main campus at Udaipur in the south and with the collaboration of SKN College of Agriculture, a branch campus, located at Jobner in the Northeast Central part of the state (See Figure 11.1). The project was jointly planned and worked out over a space of one and a half years with faculty in the agricultural sciences. Actual fieldwork was conducted through two **kharif** cropping seasons, from 1982 to 1983, in Sikar District close to Jobner. This district typifies the arid and semi-arid characteristics of rainfed agriculture in the state. While space limitations preclude an exhaustive recounting of the project, it is valuable to review the discovery and insights gained from indigenous knowledge and production techniques.

The project was designed to provide: 1) a technical and socioeconomic baseline study of the agricultural system in Sikar District that would be of use in future research programs, 2) linkages between on-station research and off-station agricultural conditions, and 3) farm level sites for adaptive trials and demonstrations under a variety of conditions. Attention was focused on pearl millet and its place in the farming systems of the fieldwork area.

The purpose of the baseline study was to provide a descriptive analysis of predominantly rainfed pearl millet agriculture to identify major agricultural features, associations, and problem domains for research planned for that region. A few years prior to the project the university had received the mandate for all agricultural research in the state, and took over research facilities and moved into regions previously under the jurisdiction of the State Department of Agriculture. This involved some reorganization and strengthening of the university's research wing that is being supported by the already mentioned national Agricultural Research Program. Due to priorities and limitations of previous research and development, dryland regions of the state had not received much attention. Sikar District, along with the other three districts that make up the Shekawati Region, fell into this category. Sikar, however, is the planned locale for a research station under the National Agricultural Research Program and this project was seen as providing a head start for the research agenda.

Although a research station was planned for the district, the program did not provide for off-station investigations. The lack of such linkages was seen as a lacuna by the scientists with whom the project was to work. Research scientists generally have few direct links to the agricultural systems for which their work is conducted. The flow of information from the field into the research process is indirect, from the state extension services to the directors of research and extension and then to the scientists. Most scientists, unless they happen to come from farm backgrounds in the same region, have little notion of agriculture as practiced off-station. Furthermore they have little opportunity or encouragement to get off-station and if they do, few contacts to approach at the farm level. The project made several contributions in this direction. First, fieldwork was conducted with a sample of farm households of differing socioeconomic characteristics spread across an agro-climatic gradient. Second, logistical support was provided for first-hand investigations and the farm contacts to do so. Third, scientists were able to observe individual problem domains not only in the uncontrolled field situation but also from an interdisciplinary perspective.

Although the university had not done research in the Shekawati region, several scientists had worked on cropping and other problems common to the Sikar area. Research on drought tolerant varieties or pearl millet (**Pennisetum americanum**) and moth bean (**Phaseolus aconitifolius**), had been conducted by scientists at Jobner in the departments of Genetics and Plant Breeding, Plant Pathology, and Entomology. The major constraints for these two crops grown extensively in Sikar are drought for pearl millet and yellow mosaic virus that inhibits plant growth and pod formation on moth beans. Experimental varieties of both crops had undergone a series of on-station and regional trials at other stations. Farm level adaptive trials were next on the agenda. In addition, control measures for a serious insect pest, white grub (main species **Holotrichia consanguinea**), had been developed at Jobner. This beetle is found in sandy regions and in its larval stage severely damages single root crops and in heavy infestations, pearly millet, sorghum, and maize as well. Since this pest is found in Sikar District a control demonstration was in order. The project's sample farmers spread across the district provided sites and the project's field staff posted in each village provided for setting up and monitoring these activities.

Sikar District is situated between 27 degrees 21 minutes and 28 degrees 12 minutes north latitude, at an average elevation of 423 meters above sea level. The district is roughly crescent shaped with an area of 2,985 square miles. The Aravalli Mountains, with peaks up to 1,000 m bisect the district on a southwest-northeast axis starting in the south central section and roughly demarcate the more arid from the semi-arid tracts. Generally the western part of the district is drier, sandier, warmer in summer, and colder in winter than areas east of the Aravallis. This forms a gradient for a number of features. In the west, soils are sandy with stabilized dunes and to the east soils grade into light loam on a plain sloping to the south. Underground water is present throughout the district but at differing depths and with a wide range of quality. The water table runs from about 150' to 30' on a rough west-east axis. Quality runs from sweet to saline on a north-south axis as one moves toward the Sambhar Basin and its large salt lake. Irrigation is concentrated in pockets, primarily in the eastern part of the district.

The climate is dominated by the June-September monsoon with average rainfall varying on a west-east gradient from 14 to 28 inches. The main cropping season is **kharif**. Winter showers are more frequent in the east where with loamier soils unirrigated crops can be taken during **rabi** season. One meteorological feature which is particularly problematic in the western part is the hot dry wind, called the **jhola**, that blows from the west near the end of the **kharif** season as the monsoon retreats down the Gangetic plain. These winds come during the grain filling stage of most crops and can dessicate a stand in a matter of days. Temperatures peak during the pre-monsoon period of May-June and are at their nadir in December-January. At Sikar weather station average maximum temperature is 43.7 C, average minimum 1.4 C, and mean temperature is 20.8 C, based on the 1978-79 data. Many areas are subject to frost.

Given the environmental conditions, agricultural productivity is generally low and erratic, particularly during the **kharif** season that is predominantly rainfed. Indicative of this are the yields for major crops during the cropping years 1977/78 to 1979/80 presented in Table 11.1.

Although temporal and spatial patterns are not indicated, the district as a whole received 235% of its average rainfall of 18.4 inches during 1977, 165% during 1978, and 64% in 1979. Too much rain damages **kharif** crops, leaches sandy soils of nutrients, damages seed beds, and/or silts over newly emerging seedlings. Too little rain results in drought. The yearly variability in yields is most striking with **kharif** crops as opposed to **rabi** crops that are grown predominantly with irrigation.

For project purposes, the district was divided into three zones on the basis of agro-climatic characteristics (See Figure 11.2). The agro-climatic gradient these zones present is illustrated in Table 11.2. The decreasing dependency on the single rainy season as one moves from Zone 1 to 3 is a function of rainfall, irrigation and also loamier soils found toward the east.

The cropping pattern across the zones shifts from almost exclusively pearl millet and pulses grown during the **kharif** season in Zone 1 to a more diversified double-cropped, irrigated regime in Zone 3 as indicated in Table 11.3. The pulses included under the monsoon crops are green gram (**Phaseolus aureus**), cowpea (**Vigna sinensis**), cluster bean (**Cyanopsis tegragonoloba**), and moth bean. The first two are consumed as lentils. Cluster bean is used for fodder although its young tender beans are used as a vegetable food and the crop is becoming an important commercial crop for the extraction of gum with several industrial purposes. Moth bean is the most drought resistant of the pulses and is found primarily west of the Aravallis as a sole crop. Moth bean is used as a lentil and also as a flour base for a deep fried vermicelli used in popular snack foods. Under rainfed conditions, however, cereals and pulses are most often grown mixed rather than as sole crops. Similarly gram, or chickpea (**Cicer arietnum**), is often intercropped with mustard during **rabi** on unirrigated land. The significant features are, however, the overwhelming preponderance of pearl millet as the **kharif** cereal under all conditions, rainfed and irrigated. Very little area is under commercial grains or horticultural crops.

Table 11.1
Sikar District
Yields (in lbs./acre) of Principal Crops, 1977/78 - 1979/80.

Year	Monsoon Kharif Crops				Winter Rabi Crops		
	Millet	Sorghum	Maize	Pulses	Wheat	Barley	Gram
77/78	100	119	84	37	1,277	1,614	862
78/79	273	272	279	41	1,283	1,432	893
79/80	165	131	43	6	1,268	1,208	671
Average	179	174	135	28	1,276	1,418	809

Source: Statistical Abstract(s), Rajasthan. Directorate of Economics and Statistics, Government of Rajasthan, 1979 & 1980.

Sikar District has a number of social features significant for agriculture and the farming systems practiced there. Historically the district was linked to the former princely state of Jaiput and until the 1950's was ruled and administered by a feudal regime that at the local level took the form of lords controlling all aspects of land and natural resource use. As with the rest of Rajasthan after Indian Independence in 1947, Sikar underwent a series of politico-administrative and land reforms into the 1960's. These reorganized and integrated the various princely states into the Indian Union, conferred land ownership on former peasant tenants, and formed elected village councils as the lowest level of democratic institutions to govern village affairs, including common lands. A concomitant change has been the demise of localized systems of subsistence production, exchange, and consumption based on patron-client structures with the inroads of commercial markets, modern transportation and closer links with wider state and national institutions (Michie 1981).

Table 11.2
Sikar District
Select Agroclimatic Features by Zone and Tehsil.

Zone	Tehsil**	Average Rainfall	% Gross Cropped Area Irrigated *	Average Cropping Intensity*	% Gross Cropped During Monsoon*
1	Fatehpur	13.8	0.3	100.17	99.7
1	Lachmangarh	23.1	1.9	101.15	98.2
2	Sikar	22.2	8.4	105.21	91.6
2	Danta Ramgarh	29.6	27.0	110.59	77.4
3	Shri Madhopur	21.9	31.9	122.96	62.2
3	Neem ka Than	28.3	29.1	123.78	52.8
1-3	All District	18.4	18.0	109.90	79.0

* Based on averages of cropping years 1977/78 through 1979/80.
** A "tehsil" is an administrative unit below the level of the district similar to a township in the US, but larger.
Source: District Statistical Handbook(s), Sikar. (In Hindi). Directorate of Economics and Statistics, Government of Rajasthan, 1980 & 1981.

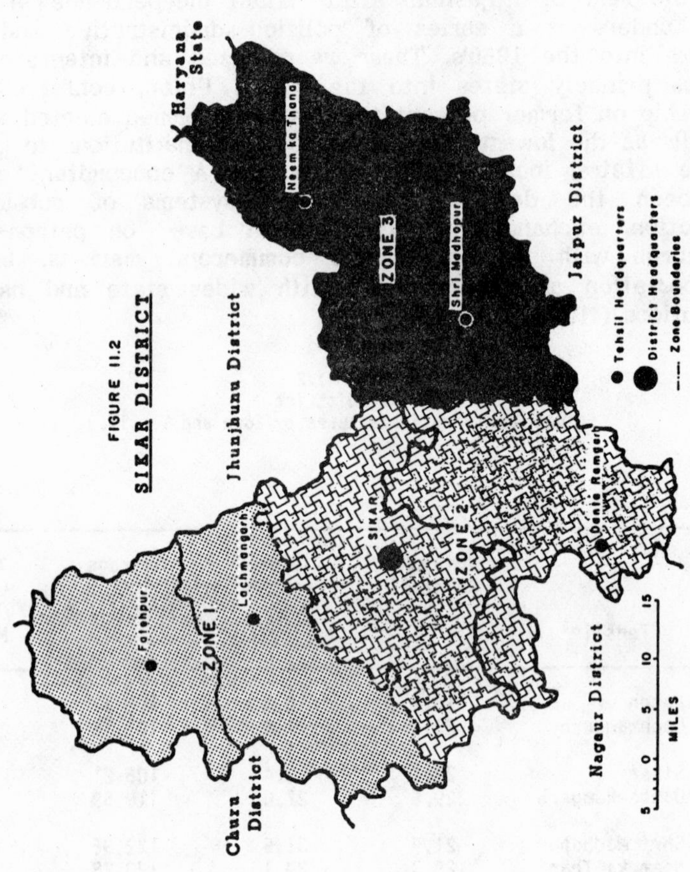

FIGURE 11.2
SIKAR DISTRICT

Table 11.3
Sikar District.
Average Gross Cropped Area and Specific Crops
as Percentage of Gross Cropped Area by Zone and Tehsil, 1977/78 - 1979/80.

			Summer Monsoon Kharif Crops					Winter Dry Rabi Crops			
Zone	Tehsil	Gross Cropped Area (Acres)	Millet (%)	Sorghum (%)	Maize (%)	Pulses (%)	Groundnut (%)	Wheat (%)	Barley (%)	Gram (%)	Mustard (%)
1	Fatehpur	18,769	66.3	---	+	33.4	+	0.1	+	+	+
1	Lachmangarh	25,080	63.6	+	---	34.6	---	1.1	0.4	0.1	0.1
2	Sikar	32,586	54.0	+	+	37.6	+	4.6	1.7	1.2	0.4
2	Danta Ramgarh	29,282	35.8	+	+	41.6	0.1	9.1	7.1	4.4	0.9
3	Shri Madhopur	38,268	32.0	3.0	+	27.2	1.2	11.5	8.4	15.3	0.9
3	Neem ka Thana	22,507	35.1	0.1	0.3	17.3	1.2	13.5	7.4	22.7	1.5
1-3	All District	166,492	46.0	0.7	0.1	32.2	0.5	7.2	4.6	7.6	0.6

Note: + equals presence but less than 0.1%.
Based on area and cropping data for years 1977/78 through 1979/80.
Source: District Statistical Handbook(s), Sikar. (In Hindi). Directorate of Economics and Statistics, Government of Rajasthan, 1980 & 1981.

Demographic changes have occurred as well. The district as a whole has had a slightly more than 100% increase in population from 1951 to 1981. In 1981 the population was 1,377,000 of which 80% was rural. Rural densities, while not high by Indian standards, run from 238 per square mile in the dry northwest to 487 in the irrigated tracts (Table 11.4).

These changes have implications for agriculture, natural resource use, and the strategies people pursue in making a living from the land. Given the fragile ecology, a decreasing **per capita** land base, and increased demand for agricultural products from both local and national markets, attempts at intensification are inevitable but also run the risk of ecological degradation. Irrigation is the most resource enhancing of intensification techniques but given water quality and depth, and capital and power constraints, it is possible only in limited areas. Expansion on previously uncultivated lands entails the loss of village common lands used for pasture, fodder, and fuelwood. In any event there is not much new land that can be brought under cultivation; 80% is already put to agricultural use and the rest is unavailable due to hilly terrain or government forest reserves (Table 11.5). Common lands are highly degraded from uncontrolled tree felling and overgrazing due to population pressure. The new socio-political arrangements on one hand have freed people from the old feudal order but have also effectively destroyed traditional institutional control at the local level over common resources.

Despite all these problematic features of the environment and the processes enumerated the situation is not hopeless. To the contrary there are some rather interesting developments that suggest solutions to many of these problems.

Desert Agroforestry

For the adaptive trials and demonstrations carried out in the project, results were rather mixed. In the first season untimely rains that came two months early upset the planting schedule, washed out seed beds, and ended in a severe drought that burned up all the pearl millet trials. The second year had good and evenly spaced rainfall that did not allow the trials to be conducted under drought conditions. For moth bean, the first year proved successful with the varieties performing near to expectation. The second year disease incidence was much heavier, forcing some re-evaluation of the resistance mechanism, whether bio-chemical or due to the hairy physiology of the

Table 11.4
Sikar District
Population, Density and Decadal Variation, 1981.

Zone	Tehsil	Population 1981			Population Density per sq. mile			% Decadal Variation 1971 - 1981		
		Rural	Urban	Total	Rural	Urban	Total	Rural	Urban	Total
1	Fatehpur	97,745	70,654	168,399	238	4,816	396	+30.51	+41.32	+34.84
1	Lachmangarh	146,913	29,215	176,128	308	9,455	367	+29.21	+31.85	+29.64
2	Sikar	211,780	102,970	314,750	368	13,675	539	+31.73	+45.05	+35.81
2	Danta Ramgarh	191,620	14,232	205,852	368	517	376	+16.91	-----	+25.59
3	Shri Madhopur	256,297	46,599	302,896	488	2,370	555	+25.30	+104.44	+33.23
3	Neem ka Thana	193,954	15,266	209,220	425	2,636	452	+31.69	+31.47	+31.68
1-3	All District	1,098,309	278,936	1,377,245	370	3,565	453	26.96	57.10	32.09

Source: District Census Handbook, Sikar. Census of India 1981. Series 18, Parts XIII-A&B, Government of Rajasthan, 1983.

Table 11.5
Sikar District.
Land Use as Average Percentage of Total Area, 1977/78 - 1979/80.

Zone	Tehsil	Forest (%)	Pasture (%)	Groves & Orchards (%)	Non-Cultivable (%)	Cultivable Waste (%)	Fallow (%)	Net Sown Area (%)
1	Fatehpur	3.9	5.6	---	3.0	0.5	20.4	66.6
1	Lachmangarh	1.3	6.3	---	2.7	0.4	17.2	72.1
2	Sikar	0.6	5.4	---	9.0	0.8	16.6	67.6
2	Danta Ramgarh	0.3	5.9	---	7.6	1.2	25.9	59.1
3	Shri Madhopur	1.3	7.1	+	9.8	2.7	13.3	65.8
3	Neem ka Thana	3.4	7.0	0.1	32.1	5.3	9.4	42.7
1-3	All District	1.7	6.2	+	10.6	1.8	17.2	62.5

Note: + equals presence but less than 0.1%.
Based on area and cropping data for years 1977/78 through 1979/80.
Source: District Statistical Handbook(s), Sikar. (In Hindi). Directorate of Economics and Statistics, Government of Rajasthan, 1980 & 1981.

varieties that vectors were learning to overcome. Similarly, the white grub control demonstration did not go as expected. The early rains in the first year completely upset the adult beetle's normal emergence and reproductive behavior, the only life phase in which they are exposed for control measures. They did not emerge from the ground in their characteristic swarms at any time. The controls, spraying of host trees, were not done. Crop damage, however, was noticeable later in the season. The same thing happened at the Jobner campus where much to everyone's chagrin many research plots were destroyed. What this underscores is the difficulty of research under erratic desert conditions where seasons are very dissimilar. A much longer time frame is necessary.

While the trials and demonstrations had their value, the most significant findings of the project came from enquiry into how farmers combine their resources and separate farm activities into a system. What was discovered was an indigenous agroforestry system incorporating trees on crop land, under rainfed conditions. The basic outlines of this system are a number of associations and interdependencies between trees, field crops, and animal husbandry that are also linked to fuel and timber concerns (Figure 11.3). Trees, crops and animals are a tripod which farmers have adapted to the fluctuations in their environment with trees serving as the stabilizing element for both crops and livestock. The spread and development of this system also allows for the intensification of production on a long term sustainable basis.

The most important tree, and the most numerous perhaps as a result of social selection, is a native mesquite (**Prosopis cineraria**) locally called the **khejri**. This tree occurs in loose to dense volunteer stands on cropped and other land. The **khejri** does well on deep sandy soils in low rainfall areas and withstands extremes of heat and cold. It grows to a height of 45 feet with spreading branches that provide light shade. Its central root penetrates up to 60 feet (Singh 1982).

A common observation by both farmers and agricultural scientists is that crop stands and yields are better under and around the **khejri** canopy than out in the open. Possible, and highly researchable, reasons for why this happens are many. The **khejri** is evergreen and leguminous with heaviest shedding of leaves during the summer before the **kharif** season. This adds organic material and nitrogen to the soil. The deep root system does not compete for moisture with crops in their root zone. Shade reduces soil temperatures and conserves moisture, both problems for crop germination in sandy soils. Later in the growth cycle this micro-environment protects against heat and moisture stresses that inhibit photosynthetic activity in crops.

The canopy acts as an umbrella, protecting pearl millet seedlings from heavy cloudbursts shortly after emergence. Pearl millet seedlings are particularly susceptible to damage at that stage; seedlings are easily beaten down and silted over by light soils during heavy rainfall. The **khejri** also provides protection against the dessicating dry winds near the end of the **kharif** season.

There are some possible problems with these associations. Trees attract birds that may damage crops although they may also attract beneficial ones. The **khejri** is a preferred host of the adult white grub beetle. The shaded micro-environment may increase soil pathogens and insects that would be neutralized by the baking action of the sun during the hot season. Mechanization of field operations is problematic since trees are volunteers and appear in random patterns. The **khejri** is slow growing and particularly susceptible to livestock grazing seedlings to the ground. Given the long maturation period it is less attractive in comparison to some exotics that have been introduced for fodder purposes.

The **khejri** is managed and harvested like a crop, the activities and sequencing of which are neatly integrated with farm operations, labor availability, animal and human needs, and natural cycles. First, the random spacing of trees presents no difficulty for animal powered operations. Draft animals, usually water buffalo and camels, are yoked singly and can plow straight tight furrows through a thick stand with ease. Second, bean like pods are produced during the hot pre-monsoon season when vegetables are scarce. The young tender beans are a supplementary, and free, vegetable, particularly for the poor. Third, harvest occurs after **kharif** is over. This creates an extra month to month and a half of employment and wage labor during an otherwise slack season, particularly where a second crop is not taken. Fourth, the **khejri** is harvested by lopping its leaf bearing branches. These are dried and the leaves beaten off. The leaves are slightly less nutritious than alfalfa and coincidentally their crude protein is highest during the cool season (Singh 1982; 272-274), the time of harvest, during November and December. Fifth, the woody loppings are used as a domestic fuel and for other purposes such as firing bricks. This frees animal dung from use as a fuel, allowing most to go into compost. Sixth, management practices dovetail with agro-meteorological cycles for the two cropping season. This is significant if a winter **rabi** crop is taken. The removal of the canopy opens the area to warmth and solar radiation during the cool winter season when it is needed.

The yearly loppings can present problems, however, due to the natural propagation of the **khejri**. Seed pods form only on

Figure 11.3

Agroforestry System Schematic

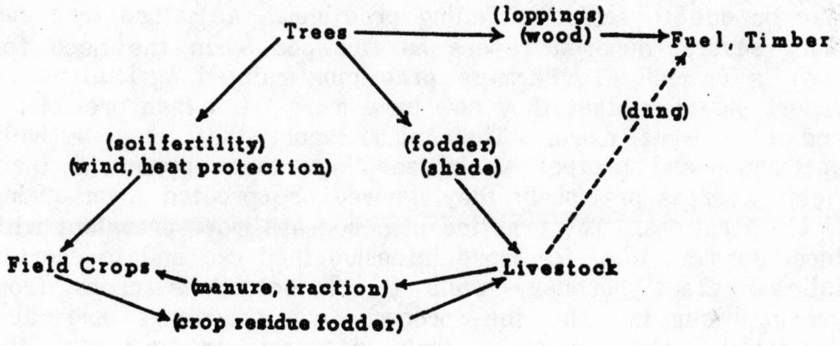

second year growth and if the tree is lopped completely year after year it will not reproduce. Most farmers do leave a few branches to allow trees to spread. In fact some farmers carefully shape their trees to make as full a canopy as possible, although this is highly variable. In any event an assured source of seed is associated with religious practices. Shrines to field gods found on most farmers' fields are always shaded by a **khejri**, which is never lopped or cut.

The associations and interdependencies found in this system are apparently long recognized by farmers. However as a dominant practice it appears to be a relatively recent phenomenon, gaining greater currency over the past thirty years. This conclusion is tentative at this writing and requires further investigation. Nevertheless there is some evidence to support this. Tree stands appear to be young, although this may be due to selective felling practices. A mature tree can bring several hundred rupees as firewood when the need for cash is immediate. Farmers practicing rainfed agriculture do report, however, that they now have more trees than previously, and they want more. They also report that they actively encourage and protect **khejri** seedlings that appear in their fields whereas previously they ignored or uprooted them during field operations. The practice also appears more prevalent with those farmers who, for more intensive land use and to shorten fallows, erect boundary walls to protect their crops from grazing animals. In the process tree seedlings are also protected from grazing that destroys or retards the reproduction of trees.

The apparent increase of agroforestry also appears linked to the social changes that have occurred over the past thirty or forty years. Population increase, institutional changes affecting natural resource use, and the linkage to wider markets all push for the intensification of production. However, the way in which agroforestry has come into its own is through a rather circuitous route. It appears to be a private response initially to the ecological degradation and/or disappearance of village common lands, the major traditional source of fuel and fodder. If sufficient quantities are no longer available from common lands then attention turns to tree resources on one's own land. Farmers also realize the direct benefits to crops and are encouraging the **khejri** on that account as well. In effect, farmers are reinterpreting and integrating trees into their rainfed farming to enhance their personal resource base and intensify the production of crops, fuel, fodder, and timber. In the process they are also decreasing the risk of degrading their own land.

Significance of Findings for Research

The significance of these findings for research on and development of rainfed agriculture should be clear. This project delineated what might be termed mega-relationships among differing farming activities in this region's farming systems that form a list of researchable problems. The findings also point to the close interdependence of these activities. The development of a package of practices would most likely have to include more than inputs and techniques relevant to the growth, management, and yield of a particular cultivar or even crop mixes. These would have to include recommendations on tree species which produce fodder, fuel and timber. The delineation of these mega-relationships does provide some guidelines and directions to that end.

First, agroforestry is a production strategy with a widespread use among the region's farmers. A research focus on this strategy would; 1) fit both the agro-climatic environment and agricultural practices already followed, and 2) most probably generate recommendations acceptable to farmers. Such a focus could be on the crop-livestock-horticultural associations already in place, as well as farm forestry, horticultural tree crops, and/or silvo-pastoralism. Some work has already been done in these directions. The studies by Singh and Lal (1969) on tree-soil relationships and by Gupta and Mohan (1982) on tree and grass combinations for silvo-pastoral systems are examples. However, such studies are few and far between. Most published research is confined by disciplinary boundaries with little analysis crossing those lines.

Second, solutions to specific problems may lie not only within one research discipline or system element but in associated disciplines and elements. As an example, drought resistance can be approached either by breeding more resistant cultivars or by research on micro-environments and their effect on physiological processes.

Third, returning to Farming Systems Methodology, the off-station environment sets the parameters within which solutions to problems are to be found. This is particularly important where targetted production systems are relatively unknown, unfamiliar, or not fully understood. The understanding of off-station parameters brings to the fore some rather crucial relationships and processes that would otherwise go unnoticed in conventional disciplinary enquiry confined to the experiment station. Conventional research sees crops as separate from animals and separate from trees. In fact most literature on the

khejri recommends that it be grown on waste and uncultivated land (e.g. Verma 1975). Furthermore the dichotomy between trees and field crops is a pervasive one that on the face of things appears to follow common sense. A good example is the otherwise excellent study presented in Gupta and Mohan's **Economics of Trees Versus Annual Crops on Marginal Agricultural Lands** (1982) with specific reference to arid lands of Rajasthan. Their methodologically sophisticated and rigorous simulation study examines numerous tree-grass combinations and juxtaposes them against field crops. While silvo-pastoralism comes out ahead in terms of monetary return, the examination of tree-crop combinations is barely considered. In fact nowhere has this author found a reference, let alone a study, of this agroforestry system as a system.

Conclusions

This case study illustrates one of the strongest assets of the FSR approach, the appreciation of locally developed agricultural practices. Numerous studies had been directed toward the problems of arid land production in Rajasthan, focusing on crops, crop combinations and infrastructural development, without realizing the potential value of the existing agroforestry techniques. In fairness, it can be noted that conceptually the system described is highly unlikely; grain crops are virtually always grown in full sun due to their physiological needs, and the scarcity of moisture would suggest that associated plants would directly compete for the little moisture available.

The techno-economic "fit" was another factor in the use of the **khejri** agroforestry combination. The farmers' needs for new sources of fuelwood and off-season food supplies contributed to the appropriateness of the technology, as did their reliance on animal rather than mechanical traction for crop cultivation. This technology, then, had been developed by the producers themselves to meet the needs of relatively poor farmers working on agriculturally marginal lands. Equally important is that the system is based on locally available low cost materials to meet needs not only for shade, but for fuel, fertilizer, and food. One of the major problems faced in agricultural development has been the differential access to improved inputs according to socioeconomic status, and this sort of agricultural technology helps overcome the unequal distribution of agricultural development benefits.

The interdisciplinary team approach to agricultural

research permits the inclusion of anthropological methodologies for the detection of existing systems which may be objects of investigation for their improvement or further propagation.

Notes

1. This discussion is based on material derived from the project "A Farming Systems Approach to Semi-Arid Agriculture in Rajasthan, India", supported by the USAID Title XII International Sorghum and Millet Program, contract no. DSAN/XII-G-1049. While the author wishes to thank the program for its support and acknowledges the invaluable contributions of others on the project team, responsibility for the views expressed herein is entirely his own.
2. Formerly the University of Udaipur.
3. An autonomous council that advises the Government of India on agricultural policy, coordinates national research and controls research funding. It acts much like the United States Department of Agriculture.

Bibliography

Chambers, R. and John Harriss. 1977.
"Comparing Twelve South Indian Villages: In Search of Practical Theory." In B.H. Farmer (ed.), **Green Revolution.** New York: Cambridge University Press.

GOI. (Government of India). 1976.
Climate and Agriculture: National Commission on Agriculture Report. Part IV. New Delhi: Ministry of Agriculture and Irrigation.

—. 1977. **Rajasthan: Rainfall and Cropping Patterns Report.** Vol. XIII. New Delhi: Ministry of Agriculture and Irrigation.

GOR. (Government of Rajasthan). 1983.
Basic Statistics, Rajasthan, 1982. Jaipur: Directorate of Economics and Statistics, Government of Rajasthan.

Gupta, R.K. and I. Prakash. 1975.
Environmental Analysis of the Thar Desert. Dehra Dun: English Book Depot.

Gupta, T. and D. Mohan. 1982.
Economics of Trees Versus Annual Crops on Marginal Lands. New Delhi: Oxford and IBH Publishing Co.

ICAR. (Indian Council of Agricultural Research). 1980.
Report of the ICAR Research Review Committee for University of Udaipur. New Delhi. ICAR.

Jodha, N.S. 1978.
"The Operating Mechanism of Desertification and Choice of Interventions". Discussion Paper #5, Economics Program, ICRISAT. Hyderabad, India: International Crops Research Institute for the Semi-Arid Tropics.

Mann, H.S. 1981.
"Management of Arid-Land Resources for Dryland and Irrigated Crops". In D.W. Goodall and R.A. Perry (eds). **Arid-Land Ecosystems: Structure, Functioning and Management.** Cambridge: Cambridge University Press.

Mencher, Joan. 1966.
"Kerala and Madras: A Comparative Study of Ecology and Social Structure". **Ethnology** Vol 5; 135-171.

Michie, Barry H. 1981.
"The Transformation of Agrarian Patron-Client Relations: An Illustration from India." **American Ethnologist** Vol 8, #1.

Morrison, Barrie M. 1979.
"The Persistent Rural Crisis in Asia: A Shift in Conception". **Pacific Affairs** Vol 52, #4; 631-646.

Nair, Kusum. 1979.
In Defense of the Irrational Peasant. Chicago: University of Chicago Press.

Singh, K.S. and P. Lal. 1969.
"Effect of 'Khejari' **Prosopis spicigera** [Linn] and 'Babool' **Acacia arabica** Trees on Soil Fertility and Profile Characteristics". **Annals of the Arid Zone.** Vol 8,#1.

Singh, R.V. 1982.
Fodder Trees of India. New Delhi: Oxford and IBH Publishing Co.

Verma, S.C. 1975.
"Forestry and Afforestation Practices." In R.K. Gupta and I. Prakash (eds). **Environmental Analysis of the Thar Desert.** Dehra Dun: English Book Depot.

12

Applying the FSR/E Approach in the Management of an AID Agroforestry Project

Edward Robins

Agroforestry is a land use strategy commonly employed by small farmers to maximize land use intensity and permit the joint production of forestry products and other agricultural or animal products. This chapter[1] reports on the Communal Afforestation Project in Rwanda where the author has been serving as Social Science Advisor to the AID mission since 1983, under the Joint Career Corps program. This innovative program permits faculty members from selected U.S. universities to serve AID as advisors for a period of 2-3 years in an overseas bureau. Thereafter, the faculty member returns to the university as a base of operations from which yet other field tours, both short and long term, ensue. The analytical know-how of university faculty is **applied** to activities in the field in this fashion; AID benefits from the expertise of individuals current in their discipline, and the individual participants and their universities gain from first-hand exposure to actual conditions in the field.

Setting

The afforestation program advanced by the joint Rwanda-USAID forestry project focuses on the introduction among the rural populace of agroforestry ideas and practices as a fundamental means of promoting reforestation and improving the farming system. Agroforestry is viewed in Rwanda as a basic intervention especially suitable on small parcels and under conditions of wood scarcity. Average farm size is just under 1 hectare; population density is the highest in Africa, at about $230/km^2$. In the project zone population density reaches $325/km^2$. Land is farmed intensively in this small, mountainous country.

Thus when an effort is made to increase tree production, it is exhibited in tree-food crop configurations. It is agroforestry more than any other feature of the project which defines the project's activities and which has called into operation a farming systems type approach to agricultural development.

At this moment Rwanda is bearing witness to a proliferation of farming systems projects. Although they do not all share the farming systems designation, in effect their approaches and goals are similar; input from farmers is solicited, trials are made on their fields, and technical packages are developed and extended on the basis of farmer-researcher collaboration. The ultimate objective is to increase agricultural production. AID has joined this effort as it begins a Farming Systems Improvement Project. In fact, the target zones are nearly identical for farming systems and afforestation projects. Thus we can speak of a distinct and concerted effort by AID to employ the FSR/E approach in the Kirambo sub-prefecture of northwest Rwanda.

There are a number of reasons why FSR/E is attractive as a means of promoting agroforestry in Rwanda.

1. There is some uncertainty about which tree species will grow best in the high altitude project zone (1800 to 2500 m). Most trials with tree-food crop interplantings have been made at lower altitudes under the aegis of the Rwanda Institute for Agricultural Research at Rubona, the Federal Republic of Germany-Rwanda Agro-Pastoral Project at Nyabisindu, and the Swiss-financed Forestry Pilot Project at Kibuye. Although research per se is not a major component of the Kirambo afforestation project, experimental trials are about to be undertaken on a demonstration/model farm set up at Project headquarters.
2. The intensive mix of crops in the Rwandan farming system is a product of many generations of trial and error with a great number of seed varieties. It is estimated that more than 350 varieties of beans are recognized in the country. There is some reluctance on the part of the farmers to introduce into this refined system new farming practices, such as agroforestry, in the absence of a convincing demonstration of their effectiveness. For this reason farmers are included in the process of identifying constraints to production; their involvement in the Project will make them more receptive to proposed interventions. These latter are a product of the assessments farmers make of their own production systems. By collaborating with farmers to improve the

farming system **in terms of farmer perceptions of its failings,** the Project hopes to introduce practices beneficial but as yet unproven in the Project zone.

3. There is a feature of the extension system as it now operates in much of Rwanda which **decreases** the likelihood that farmers will willingly try technologies proposed by interventionists. Extension agents are usually some years younger than farmers and considerably less experienced at farming. Moreover, the agents are not highly trained professionals whom the farmers can respect; quite the contrary. They are mostly primary school graduates, some with a year or two of secondary school, who know some things about farming but even less about most other things and whom farmers do not hold in particularly high regard. The issue is: how to strengthen this absolutely essential link between the project and the farmer given the lack of confidence farmers have in extension agents. The project is training agents in the agroforestry approach and attempting to persuade them to seek first the opinions of farmers and only afterward to provide counsel. In working to strengthen the extension program the project is striving to develop the mechanism by which the idea of collaboration can be realized.

The increased presence of trees on the farm rather than in communal forests should appeal to Rwandan farmers on economic grounds. Rwandan farm families have weak purchasing power; agroforestry requires a minimum financial investment, and the proximity of wood to the **rugo** (living area) reduces transportation and labor costs. Agroforestry is highly touted in Rwanda at present as the unique solution to a unique problem (Egli and Raquet 1985; Baumer 1985; Nair 1983).

This paper will discuss the effort being made by the communal Afforestation Project to introduce agroforestry practices into a three commune area by employing elements of FSR/E. This approach is characterized by a three-fold program: 1) soliciting from farmers descriptions of the tree component in their farming system; 2) conducting on-station and on-farm trials with species which are of proven agroforestry value (although not all of them well understood in the Project zone); and 3) training extension agents to collaborate with farmers as a means of promoting cooperation between farmers and the project.

Project Site

The project is located in the three commune area which constitutes the sub-prefecture of Kirambo, prefecture of Ruhengeri, in northwest Rwanda. The population of this zone, according to the 1983 census, is 134,233 persons. The total area of the zone is 492 km^2. Population density is highest in Nyamugali Commune, at 323/km^2, followed by Butaro at 301/km^2, and Cyeru, at 222/km^2.

The region consists largely of red and dark **foncee** schist soils of high quality, cultivable to depths of 50 to 80 cm. Land is farmed on slopes of up to 60%. The soils are said to be of good structure, texture and color (**Rapport Annuel 1984**).

The principal crops in this region are beans, bananas, maize, sorghum, peas, potatoes, and sweet potatoes. Some livestock is raised, most notably cattle, goats, sheep and chickens. Trees tend to be planted on the crests of hills and on land too steep to farm. Valley bottom land is the richest and is never planted in trees. Small woodlots are occasionally found on the farms, although increasingly the competition for agricultural land makes this practice impractical.

The Afforestation/Agroforestry Project

The four-year project began in April 1984, in Cyeru Commune, with a $500,000 grant from AID and a Government of Rwanda (GOR) contribution monetized at $335,000. At that time, the first rainy season of the year (March-May) was coming to an end and the project established as its immediate goals the repair of poorly functioning sectoral nurseries and the raising of seedlings for the next season (September-December) when trees would be planted. Some 20,000 seedlings per nursery were raised by October, and over 300,000 seedlings were planted in the commune. The most popular species were Eucalyptus, Pine, Cyprus, and Grevillea. Only the last is an agroforestry species. Most planting was done in communal forests, of which 140 ha. were planted. An additional 30 ha were planted along roads. The project estimates that 90 ha of individually owned land were planted, mostly in woodlots. Some agroforestry planting took place during the first season, although the project's initial efforts were not in that direction. There were farmers who, having spoken with communal agricultural officers, already knew which trees could be planted successfully in fields, along contour lines, on boundaries, and even among other crops, and so some plantings were made in these configurations using the species **Grevillea robusta** and **Markhamia platycalyx.**

The project began activities in a second commune, Butaro, in January 1984, and for the first season likewise directed its effort at planting communal woodlots. Concentrating on this activity at the start when nurseries and extension agents both are new to the project has enabled the project to satisfy targets set in the project paper for communal woodlot planting (a total of 400 ha in three communes over four years). It is a political expedient as well to concentrate first on communal land. The **bourgmestre**, the administrator of the commune, is a political figure whose support for project activities needs to be curried. The commune derives revenue from the sale of wood cut in communal forests. By taking care of its communal "obligation" first the project is more likely to benefit (and in fact already has) from the **bourgmestre**'s support. In particular, the **bourgmestre** allocates **umuganda** (communal) labor; one member of each household is obliged to spend one half-day per week in non-remunerated work for the good of the commune. The project has profited from **umuganda** repair efforts on roads leading to nurseries, although **umuganda** used in the nurseries themselves has yielded low productivity. The target for Butaro Commune this year is the planting of 90 ha of communal forest.

Since the activities of extension agents in Butaro Commune have been related mostly to developing nurseries and organizing the reforestation of plantations, the training they have received has been directed toward these activities. In Cyeru Commune, where the agroforestry program in underway, the agents are receiving additional training in agroforestry theory and practice based on the module developed by Swiss technical assistance. This module uses visual aids to portray various species in their altitudinal zones, illustrating the uses to which the species can be put. Agents are instructed in how to make presentations to individuals and groups of farmers. The agents practice the routine in a workshop; in the field their activities are overseen by extension officers. Officers are responsible for helping agents to refine their technical know-how and communications skills. The extension program for the first six months of 1985 increasingly has been oriented toward agroforestry. The entire cadre of agents is yet to participate in a newly created bi-weekly program designed to further improve the communication skills suitable for work in rural Rwanda. The content of the program will be developed by the AID Social Science Advisor in collaboration with a Rwandan communal agricultural officer and a Rwanda extension supervisor.

This year too marks marks a change in the composition of species in the nurseries. The proportion of non-agroforestry species has been reduced in favor of agroforestry varieties. At present, about 15% of nursery stock is in agroforestry species, with an additional 5% in fruit trees. These percentages will be increased in the coming seasons. Approximately 60% of the woodlot species is devoted to **Eucalyptus.** This species, overwhelmingly preferred by rural Rwandans for its rapid growth, straight wood and mosquito repelling properties when burned, is nonetheless counter-productive as an erosion control measure and toxic to food crops growing in its vicinity. The project will try to wean farmers of their heavy dependency upon **Eucalyptus** in favor of some of the faster growing agroforestry varieties (such as **Grevillea** or **Cedrela**).

The agroforestry species which are being grown at this time are noted below, with their project uses and dominant features;

Tree Species	Special Qualities, Uses in Agroforestry
Acacia mearnsii	nitrogen fixation, fuel, poles, mulch
Albizia gummifera	local species, nitrogen fixation, fodder, mulch
Calliandra calothyrsus	nitrogen fixation, fuelwood, fodder, mulch
Casuarina spp.	nitrogen fixation, windbreak, fuel
Cedrela spp.	fast growing, wood, mulch, light shade, fuel
Grevillea robusta	fast growing, wood, leaf litter, fuel
Leucaena leucocephala	nitrogen fixation, fodder, fuel, anti-erosion terracing, mulch
Markhamia platycalyx	poles, wood, light shade, leaf litter
Sesbania sp.	local species, fast growing nitrogen fixation, light shade, fuel, fodder
Grass Species	
Pennisetum purpureum	anti-erosion, fodder, bean stakes, roof supports
Setaria spp.	anti-erosion, fodder
Tripsacum spp.	anti-erosion, fodder
Shrub Species	
Crotalaria spp.	legume, fodder, green manure
Desmodium spp.	legume, fodder, green manure
Tagetes spp. (marigold)	green manure, insecticide
Tephosia spp.	green manure, insecticide

To date **Grevillea** and **Cedrela** are the exotic species which have done especially well in the project zone. The growth of **Leucaena** in the nurseries has been disappointing; an inoculant will be used in the coming season to see if growth improves. **Leucaena** is among those species with which the project will experiment; trials with **Leucaena** in Rwanda have not been conducted above 1800 m. Trials will likewise be conducted with a number of high altitude species having done well in Kenya, such as **Tipuana tipu** and **Calliandra calothyrsus**, which have yet to be extensively tested in Rwanda.

It is expected that hedgerow species will be strongly promoted in the coming year. Prunings can regularly be taken for fuelwood and in conjunction with **Setaria** grass the hedgerow is an effective erosion barrier. The project will experiment with **Abellia caffre** in this configuration.

A final activity planned for 1985 is the development of a demonstration "farm". A one ha parcel has been given to the project by the sub-Prefecture and will serve to illustrate integrated agroforestry, agriculture and animal husbandry. Traditional crops will be grown, such as beans, bananas, Irish and sweet potatoes, sorghum, maize, peas and wheat. Agroforestry species known to do well in the high altitude zone will be planted in addition to those species which are as yet unknowns in the area. These latter will be introduced on a trial basis. Fodder grasses will be used for erosion control and as feed for 2 milk cows, to be purchased with the receipts from the sale of the first season's harvest. Thereafter it is expected that the sale of milk and other agricultural products will provide income to finance the farm's operation.

Agroforestry as FSR/E

The agroforestry component of the project follows an approach which seeks to involve farmers in the identification of constraints to agricultural production (in this case with regard to wood) and in on-farm attempts to eliminate them or render them less constraining. The use of extension agents to link project management and farmers is fundamental to successfully realizing this goal. Although research, and especially the development of new varieties, is not a major component of the project, experimentation with untried varieties will be undertaken, as already noted. (The Farming Systems Improvement Project which is shortly to begin in the Project zone includes a major research component.)

The objectives of the project are derived from the reforestation program established for the country as a whole. The principal goals of the program are to increase the supply of fuelwood and lumber, and to check erosion. In 1981, the rate of reforestation in Rwanda was 8%, or .04 ha per person (by comparison, the rate of sub-Saharan Africa is 24%, or 2.4 ha per person). Rwanda's annual wood needs are 5 to 6 million m^3. At present forests produce .9 million m^3. As 10 m^3/ha/year can be produced on the average, 400,000 to 500,000 additional ha must be planted. A family of 8 persons needs 2.2m^3 of wood, 70% of which is used as firewood. Integrated agroforestry can yield 7 m^3/ha/year, more than enough to meet Rwanda's growing demand (**Vulgarisation Forestiere Communale** 1981; Kalinganire and Egli 1985). The contribution of the project to this effort will be realized in planting 400 ha of communal forest and 2500 ha of individually owned land. On-farm plantings will be made in woodlots, on field boundaries, along roads, as windbreaks, and as tree-foodcrop intercroppings. A nursery will be established in each of the 12 or so sectors of each commune, providing seedlings free-of-charge to farmers. One extension agent per sector manages the nurseries and is charged with extending agroforestry technology to farmers.

Diagnosis and Design

Raintree (1983b;173) states that the design of a program for agroforestry interventions must be based first upon a diagnosis of the farming system. By involving farmers in this process, it is expected that they will take a greater interest in the program of interventions which follow;

> ... no agroforestry technology, no matter how technically correct or elegant, will have a significant impact on the landuse scene unless it is adopted by a significant percentage of intended landusers.

This approach is used at the International Center for Research in Agroforestry (ICRAF) in Nairobi and serves as a model for the project (Raintree 1984, 1983a,b; Rocheleau and van den Holk 1984; Torres and Raintree 1984; Hoekstra 1984). Diagnosis has been undertaken by implementing a socio-technical survey in order to identify tree-growing and wood-use practices followed by farmers in the project zone. The information sought includes: trees grown on the farm and their location;

uses of various tree species; wood preferences; access to wood, live and dead, by use, location and family member; wood sales and purchases; wood-growing practices, including tree-food crop interplanting; energy requirements; and experience with and assessment of the extension service. The survey is not an attempt to record all features of the farming system/rural family life. Rather, the emphasis has been placed upon trees and how they fit into the farming practices of the family, and resource management practices, needs and potentials. It is not likely that recommendations domains will be identified. The FSR/E project will include an extensive diagnosis of local farming systems as a prelude to identifying recommendation domains. The interventions planned for the agroforestry project will be considerably fewer than those for FSR/E, and less complex as well. Only a limited number of species will likely do well in the high altitude project zone, where the crops grown on one farm are nearly identical to those grown on neighboring ones. The packages proposed to farmers will be basically of two types; hedgerow-fodder grass mixes, largely for erosion control, such as **Abellia caffre** and Setaria grass; and tree-food crop mixes, for soil improvement, erosion control, fuel, fodder and wood, using such species as **Grevillea** and **Cedrela** (the latter was previously untried in the area but proved to adapt well in the first year of project operation). From the point of view of project management, a more extensive survey would be impractical at this time, both financially and in terms of available human resources; technically it is not necessary either, given the limited number of unknowns with which the project is confronted. The survey that was conducted was limited in scope and duration, and focused on a few key areas; these will guide project personnel in counseling farmers based upon the identification of their particular wood-use needs and crop configuration/land availability. Too often in surveys of this type data is collected with unbridled enthusiasm, for the sake of "comprehensiveness" (and because of access to a computer), and subsequently constitutes a mass of material difficult to manage and interpret. In this regard anthropologists have an important role to play. Having relied so heavily on the qualitative approach, anthropologists now first coming to quantified analysis are sensitive to its shortcomings, particularly the constraints placed by the questionnaire format on responses; statistical analysis, with which development agencies are enamored, is so regularly misused as to produce a pool of misinformation detrimental to project development. It seems that statisticians themselves share the anthropologists' anxiety concerning this abuse. The skillful blend of qualitative and quantitative analysis of which anthropologists are capable would promote more accurate identification of indigenous views and practices. The survey data for the reforestation project in fact will be analyzed using an IBM/PC and an SPSS program; yet other information was gathered previously by interviewing farmers/farm families (in fact largely as a basis for formulating the survey questions). The anthropologist makes a special contribution by doing the field interviews and afterward designing the questionnaire him/herself.

Fundamental interventions, simple and effective, often can be proposed after consulting with farmers and soliciting their points of view. Farmers are capable of articulating problems and production constraints if given the opportunity to do so in a dialogue with fieldworkers. In this sense extensive surveys often are undertaken to justify an approach which could just as easily be designed using qualitative methods; but the "subjective" interpretations resulting from the efforts of individual fieldworkers conducting interviews are generally viewed as less reliable than the "objective" statements generated by quantified analysis. The fact that the personal interpretations of the researcher figure as prominently in the latter methodology as in the former often is unappreciated. The qualitative approach gives the appearance of being less scientific and the interventions derived from it harder to justify and defend to individuals representing the agencies which fund development projects. Nonetheless its value is considerable. As social science advisor, I regularly ask contractors to identify their objectives, and the information they need to reach them, whenever they come to me with a request for a survey. Often they don't understand the relationship between the information sought and the question asked. Knowing what not to ask, what not to collect, is important, too.

The socio-technical survey is the most extensive data collection effort undertaken by the project, but not the first. Earlier surveys based on field interviews yielded useful information for the first training session of monitors. It was established that farmers were well aware of erosion, of the need for reforestation, of decreasing soil fertility. They recognized shortages in fuelwood and lumber, but did not view trees as sources of fodder and mulch. In the training of extension agents these generalities were brought out and constituted a base from which the agents could begin their work. The anthropologist collecting such information and presenting it to extension agents afterward makes both an analytical and implementational contribution to the FSR/E type program.

Collinson (1984;5) has proposed that the role of on-farm

research is to "target research output to the needs of farmers", but not to increase the "number of research questions". The pitfall here, he continues, is that;

> ... research and extension establishments are dominated by technical perspectives and criteria, whereas farmers make decisions from a managerial standpoint. Technical compromises are essential to good farm management, but are often seen as farmer shortcomings; conversely, optimal technical packages may well be managerially inappropriate. If research and extension are to offer useful recommendations to farmers, they must look at the whole farming system within which a farmer makes his decisions (ibid;4).

Technical advisors can be insensitive to the non-technical considerations which preoccupy farmers, and as a result may be critical of the project's development in the absence of knowing the non-technical factors of which farmers' decisions are made.

The survey just completed for the project will serve as a vehicle through which project personnel will learn more about the farming system, and at the same time will inform farmers of those aspects of the farming system which the project has identified as critical, or potentially critical, to increased production. The survey will serve also as a means of educating farmers, as a means of extending to them information which technicians, based upon their own expertise, have identified as significant. This "give and take" is the dialogue which FSR/E strives to induce, where each side learns from the other and works together for the good of the farm family.

Research

The model agroforestry demonstration plot at project headquarters at Kirambo will "experiment" with a number of varieties as yet unproven in the project zone, including **Leucaena, Tipuana**, and **Cedrela**, as well as varieties indigenous to the area - **Sesbania, Albizia** and **Markhamia.** The actual research component of the project will be linked to on-going research at the **Projet Agro-Pastoral** at Nyabisindu, a Federal Republic of Germany funded and operated research station, the Swiss financed **Projet Pilote Forestiere** at Kibuye and the agroforestry division at ISAR, the national agricultural research institute (**Institute des Sciences Agronomiques de Recherches**). The Nyabisindu project, begun in 1969, is regarded as the most successful of its kind in Rwanda, and is the model upon which other integrated silvo-agricultural projects have been based. The program includes research, extension, and on-farm trials for promoting integrated forestry. The project is the country's principal source of forestry and agroforestry information and counsel (**Projet Agro-Pastoral de Nyabisindu** 1983;18);

> Soil conservation is the starting point for intensifying agriculture. To achieve it, the project prefers biological means; anti-erosion lines set perpendicular to the slope of the hill with shrubs and trees which protect the soil in different ways and, at the same time, furnish fodder for livestock and material for mulching, fruit, and wood. (Author's translation).

The need for coordination between the extension and research programs is appreciated in principle in Rwanda (Egil 1985), but is only slowly materializing in fact. The results of trials with agroforestry species are regularly reported by ISAR and the Nyabisindu and Kibuye projects, but the publication of a planned **Revue Forestiere Rwandaise** has yet to occur. Moreover, the extension materials which must regularly be improved with feedback from the results of on-farm trials and new research still are not sufficiently diffused in the rural environment. As many Rwandan researchers are not accustomed to working in the field, it may be some time yet before the necessary coordination is realized.

In the meantime, farmers who are following sound agroforestry practices need to be encouraged. Agroforestry is in fact an old practice in Rwanda although its scientific application is new. In the past, forests and pasture lands were mixed, live fences were used, hedges were grown with **Erythrina abyssinica** to protect crops from livestock, and **Markhamia**, **Acacia**, and **Albizia** were planted to check erosion (Munyarugerero 1985). It is not likely that farmers knew at that time about the nitrogen fixing properties of these species. The current problem in introducing agroforestry is less one of promoting a novel idea than it is of the on-farm division of labor. Generally in Rwanda men care for trees and women work in fields. Men often are unaware of management needs in fields to combat erosion and improve deteriorated soils. Men rarely involve themselves in obtaining firewood, and are less cognizant of the need to look increasingly further afield from the farm for a supply. Although decreasing agricultural production has required that men and women work more closely than before to better manage the farming system, the basic division of responsibilities on the farm necessitates that the extension effort reach and educate women and men together. Despite the familiarity of agroforesty in the Rwandan context, most rural dwellers know little about the principles of good tree-foodcrop management. Rwanda farmers will likely be receptive to agroforestry as they see convincing demonstrations of its value. Already a number of species are found with food crops along the contour lines of hills - **Markhamia, Ficus** spp., and **Euphorbia** spp. If the project can accentuate positive practices while reassuring farmers that the unpleasant experiences with coffee will not be repeated, then agroforestry should find a welcome home in Rwanda. (Coffee tree monoculture is obligatory on the newly established farms in eastern Rwanda; some land has been taken out of crop production as a result.)

Conclusion

The FSR/E approach to development is particularly attractive to anthropologists for its insistence on involving the practicioners of indigenous agriculture in the design and implementation of projects. Anthropologists long have championed the value of using local perspectives as a basis for understanding socio-cultural systems and increasing our knowledge of the ways in which humans can successfully exploit the multitude of habitats in which they live. FSR/E gives explicit recognition to the need for having farmers identify agricultural production constraints to researchers, and subsequently for including these farmers in the lengthy process of experimentation with new technologies and configurations as a means both of benefiting from their considerable agricultural expertise and of fully involving them in the effort to improve the farming system of which they are an essential component.

There is much for the anthropologist to contribute to the FSR/E effort. In the initial diagnostic phase anthropologists employ their skill at interviewing to obtain a sense of the important issues which will be more completely investigated in the extensive survey. A qualitative methodology is highly valued at this time in the development of the project, and the anthropologist is the scientist most capable of employing it successfully. Anthropologists who employ quantitative methods in their analysis are more valuable yet for the coordination they can promote between the two methodologies of informal and formal surveys.

As the project unfolds, the anthropologist continues to

mediate between the various participants - scientists of different fields and persons of different cultural backgrounds; institutions in the host country, those in Washington, and at the university. The emphasis in FSR/E upon collaboration across scientific disciplines and cultures makes the field particularly suitable for anthropologists who, by training and experience, possess the expertise to identify the cultural bases of conflict.

The anthropologist who manages projects for AID is a more unusual species yet. The implementation of projects is a true test of applied anthropology. Although it is customary to find social scientists participating in the design of projects, those that are involved in implementation as well gain an appreciation of the extent to which the design is in fact implementable. At the same time, project management enables the anthropologist to put into practice in the cross cultural context values derived from the study of culture. It is a worthwhile exercise for the anthropologist to practice what has been promoted in the abstract.

The afforestation project in Rwanda has found FSR/E to be a useful means of bringing agroforestry to the rural populace. In the densely populated regions of northwest Rwanda the intensity of farming is such that little land is available for woodlots. As wood is the principal source of cooking fuel for the country, and is used as well for carpentry and construction, the scarcity of land available for woodlots is constraining. Trees which can be planted among other crops can help to solve the problem of wood scarcity while not taking land out of crop production. Equally important, however, trees, shrubs and grasses can serve as barriers to erosion and help to promote increased soil fertility. Nitrogen fixing species contribute even further to this process. Some species provide food for human consumption, in the form of fruit, and fodder for livestock. In principle the bounty agroforestry can furnish to the Rwandan farm and farm family is impressive.

In order to make agroforestry a reality the project has faced three fundamental constraints which the FSR/E approach addresses and can help to overcome. First, there are agroforestry species known to be effective in Rwanda but at altitudes lower than those in the project zone. Through on-farm and demonstration trials the project will experiment with these varieties to identify those which do well in the high altitude zone and simultaneously serve the needs of the area's farm families.

Secondly, the Rwandan farming system is a complex product of generations of trials with numerous varieties of the many crops grown on each of the intensively farmed hills in the country. Farmers are unwilling to introduce new components into their systems when these are imposed from the outside. By involving farmers in the identification of constraints and the ensuing dialogue on how to remove them, the project obtains both their cooperation and their considerable knowledge of the Rwandan agricultural setting.

Finally, the means by which dialogue is promoted and maintained is the purview of the extension program. Extension agents have the most regular contact with farmers but usually know less about farming than do farmers themselves; the latter are understandably reluctant to take too seriously the counsel of extension agents. The project, by providing agents with training in agroforestry theory and practice and communications skills, strives to make the agent both more knowledgeable about the activities he/she is promoting and more receptive to what farmers have to say. A good rapport between agents and farmers takes time to build, but it is essential for the type of collaboration FSR/E requires.

The Rwandan Communal Afforestation project will continue to develop and refine its methodology in the next two years, and will continue to monitor the progress it makes in promoting agroforestry. Already there is recognition among farmers of wood shortages and decreased soil fertility; the idea of agroforestry is not entirely new and it appeals to them. Through a collaborative effort the project hopes to obtain the support of farmers in trying new methods, new configurations, and new species as a means of increasing the number of trees planted on the farm. Managing trees as crops is a long-term affair; patience and a long-term commitment is needed. If farmer interest can be maintained through the initial phases of this effort, the demonstrations of success should be sufficiently convincing later on to enlist the support of the majority of the inhabitants of Rwanda's rural regions.

Notes

1. This paper presents the views of the author and not necessarily those of the US Agency for International Development.

Bibliography

Baumer, M. 1985.
"L'Agroforesterie Peut-Elle Aider au Developpement du Rwanda?" In **Compte-Rendu, Journee d'Etudes Forestieres et Agroforestieres**; 89-102. Rubona: ISAR.

Collinson, M. 1984.
"Commentary". In **Africa Workshop on Extension and Research.** Washington, DC: World Bank.

Egli, A. 1985. "L'Avenir Agroforestier au Rwanda - Quelques Possibilites d'Intervention." In **Compte-Rendu, Journee d'Etudes Forestieres et Agroforestieres**; 159-179. Rubona: ISAR.

Egli, A. and K. Raquet 1985.
"L'Agroforesterie au Service de la Fertilization et de la Conservation du Sol au Rwanda." Butare: ISAR.

Hoekstra, D., F. Torres, J.B. Raintree, T. Darnhofer, and E. Kariuki. 1984.
Agroforestry Systems for the Semi-Arid Areas of Machakos District, Kenya. Working Paper #19. Nairobi, Kenya: ICRAF.

Kalinganire, A. and A. Egli 1985.
"Le Role des Arbres en Milieu Rural au Rwanda: Resultats d'une Enquete Menee en 1983". In **Journee d'Etudes Forestieres et Agroforestieres**; 139 - 153. Rubona: ISAR.

Munyarugerero, G. 1985.
"Les Systemes Agroforestiers Traditionnel au Rwanda, Quelques Example." In **Journee d'Etudes Forestieres et Agroforestieres**; 103-110. Rubona: ISAR.

Nair, P.K.R. 1983.
Some Promising Agroforestry Techniques for Hilly and Semi-Arid Regions of Rwanda. Nairobi, Kenya: ICRAF.

Projet Agro-Pastoral de Nyabisindu. 1983.
Rapport Annuel. Nyabisindu, Rwanda: Republique Federale D'Allemagne.

Raintree, J.B. 1983a.
Guidelines for Agroforestry Diagnosis and Design. Working Paper #6. Nairobi, Kenya: ICRAF.

———. 1983b. "Strategies for Enhancing the Adoptability of Agroforestry Innovations". In **Agroforestry Systems** 1;173 - 187. The Hague: Martinus Nijhoff.

———. 1983c. **The Agroforestry Approach to Land Development: Potentials and Constraints.** Nairobi, Kenya: ICRAF.

———. 1984. **A Systems Approach to Agroforestry Diagnosis and Design; ICRAF's Experience with an Interdisciplinary Methodology.** Nairobi, Kenya: ICRAF.

Torres, F., J.B. Raintree, M.V. Delmacio and T. Darnhofer. 1984. **Agroforestry Systems for Smallholder Upland Farmers in a Land Reform Area of the Philippines: The Tabango Case Study.** Working Paper #18. Nairobi, Kenya: ICRAF.

13

Social Science in FSR: Conclusions and Future Directions

Ben J. Wallace
and Jeffrey R. Jones

The development of FSR has come on the heels of dramatic changes in world agriculture. During the past 100 years major advances have been made in farm mechanization, followed by advances in the use of chemicals to enhance soil fertility, breeding techniques to improve crop yields and plant growth characteristics, and the discovery of a broad range of agrochemicals for the control of insects, bacteria and fungus which previously had posed major problems to production. In the United States, the institutional relationships between research universities, extension services and farmers developed throughout the century of change in a gradual adaptation to the technological capacities of researchers and the needs of farmers. The result was the creation of an oft-cited technology development and transfer system, although there is an increasing awareness that the system also depends on US social conditions to properly function. Specifically, the education of farm children in agricultural research universities, and the social status relations between farmers and extension agents have been recognized as crucial factors (Whyte and Boynton 1985). Major efforts have been made to improve the agricultural research capabilities in developing countries through the establishment of the International Agricultural Research Centers, and through national level institutions with responsibilities for research, extension, or both.

The Farming Systems Research approach to agricultural development grew out of the recognition that biological research must be integrated into local production contexts, as it was in the US through the educational system. As evidenced by the contents of this volume, FSR has begun to bridge the gap between on-station and off-station conditions, and now faces the major challenge of efficiently implementing new technologies without causing ecological imbalances in relatively new production environments, and social imbalances in older areas.

It should be clear from the contributions to this volume that FSR is not a monolithic approach. Variations in terminology are evident, as are differences in concepts as basic as multidisciplinary team formation. While this may smack of theoretical impurity in a purely conceptual sense, it accurately reflects the diversity of conditions facing agricultural development. A more important question is to what extent the heterogeneity in the application of FSR leads to project implementations which more effectively achieve development goals. This should be a basic criterion in considering the cases described in this volume.

Scientific Revolution and Normal Science in Agricultural Development

Agricultural development over the past century can be viewed within the framework suggested by Thomas Kuhn for describing scientific development (1962), in that a revolutionary paradigm for agricultural production has been introduced which views all production components as malleable, and many to be artificially reproducible. Agricultural technology has been undergoing a period of "revolutionary" development over the past decades, with the introduction of new inputs for modifying soil and ecological conditions, and new capabilities for modifying crop genetic make-up, reflecting new production theories. An important point to make in using this analogy for agricultural development is to note that the theories identified have not been exclusive and successive, but simultaneous and complementary, so their effects are interactive. In developing countries especially, the rapid succession of non-exclusive theories has left no opportunity to settle into a period of "normal science", where the paradigm and its component theories can be carefully studied and tested to determine its more universal applicability.

Following Kuhn's framework, the objective of FSR is to proceed with "normal science" activities in agricultural development. This activity requires the implementation of new agricultural "theories" in a broad range of conditions to determine their universal validity. The outcome of this "normal science" activity is the generation of corollaries to major theories to explain their operation under the observed conditions. In the US, normal science is carried out to a large extent by farmers or farmers' children; social and institutional networks provide farmers with access to research station results, either through their own efforts or the education of their children. In developing countries these same "normal science" relationships have developed between research institutions and principally wealthy farmers. The development of FSR responds to the need to make research useful and available to a broader spectrum of agricultural producers to achieve an ample base of agricultural development for social equity and to support national economies. Research and extension efforts in developing countries must be seen as investments whose value is judged by the amount of wealth and social well-being generated, and FSR provides a crucial step in ensuring that the results of these investments are made available in a useful form to farmers.

One outcome of the most recent period of scientific revolution in agricultural development has been a change in ecological relations. When agricultural economies are left to their own devices, equilibria often develop wherein social activities correct disequilibrating aspects of ecological processes associated with the agricultural economy. New Guinean slash and burn farmers practice periodic ritual pig hunts which coincide with increased pressure by the pig population on the poorly protected slash and burn plots in the forest (Rappaport 1968). Among Zinacanteco Mexican peasants, social obligations on successful individuals drain off wealth and time, thereby avoiding a breakdown of the local socioeconomic equilibrium in the village (Cancian 1965). While it would be unrealistic to assert that all traditional agricultural societies tend to find an equilibrium, the socially defined controls of existing populations and the technological and physical limitations tend to slow disequilibrating change. These semi-equilibria are broken down by the combined effects of market forces and new technologies. In notable cases traditional societies have reacted to disequilibrating factors through the construction of alternative buffer mechanisms when exogenous factors accelerate change, but in much of the world the forces for rapid change have overwhelmed traditional systems. This "resistance to change" among farmers has been a topic of social and psychological investigation, but may in fact be based on a rational evaluation by the affected farmers of potential impacts on their own well-being (Jones 1976). Given the interrelation of social, economic and biological factors, a more detailed understanding of local conditions is a prerequisite for proposing research or agricultural improvement strategies which will not create more problems than they solve.

Historically, technological change has paralleled population growth, and increased pressure on land resources has been accompanied by increased agricultural productivity through the mechanism of increased investment. Patterns of land use intensify, e.g. from long fallow to short fallow, or from single cropping to double cropping, as population pressure increases. The increased application of resources, especially labor, has been a basic process for meeting growing demands on an inelastic land resource base (Boserup 1967).

In recent years, the expansion of agricultural production has been accompanied by the development of agrochemicals, specifically herbicides, which have significantly altered the relations between man and land. The humid tropical areas of the world have been little utilized because of difficulties of management, especially the problems of brush clearing and weed control. Fallowing is used in these areas as much for the elimination of weeds as it is for the restoration of fertility (see for example Carter 1969). Herbicides represent a dramatic improvement over "mechanical" weeding, and there has been a corresponding increase in their use; in Panama 54% of all agrochemicals imported between 1970 and 1978 were herbicides, and over the period 1960 to 1984, agrochemical use in general increased from less than 1,000 to nearly 8,000 metric tons (Espinosa 1985). Since weed control is a major limitation in the cultivation of tropical lands, the use of herbicides has caused significant changes in the productivity equations for these areas. The result has been a rapid and widespread change in the intensity of humid tropical land use all over the world, whose most visible manifestation is deforestation.

Recent interest in fuelwood production is a response to the increased capacity for clearing land, and keeping it cleared, in developing countries. For example, fuelwood is used for heating and cooking in the majority of Central American households, but for rural consumers it was essentially a free good; fallow plots and local forests were ready sources, and were not far from home until fairly recently. As land-use intensity increases, fallow plots and forests tend to be located further from households. The same farmers who cleared the land are themselves surprised by the completeness of the deforestation that has occurred, and some have felt the need to reforest for fuelwood and other needs on their farms (Jones and Price 1985).

The scientific revolution in agriculture introduced a new set of social and biological conditions. As the new production paradigm is applied under different conditions, unforeseen consequences arise, such as social inequalities and ecological perturbations, in addition to increased food production. FSR is one mechanisms which is now being implemented to develop solutions to the problems encountered. While FSR activities have drawn social scientists into a more intimate collaboration with agricultural development projects, at the same time there has been a development of tensions within projects as a result of interdisciplinary involvement. These experiences suggest future directions for social science participation in agricultural development implementation which are of both theoretical and immediate interest.

Social Science Contributions to FSR

Social scientists have been involved in nearly all phases of FSR activities, from **sondeos** and characterizations to the implementation of tested technologies. The most common activity, however, is in the initial phase of research to help in the generation of baseline data and the discovery of indigenous technology management patterns.

The characterization phase permits the widest application of social science techniques to FSR. Since this phase is to a large extent exploratory, an open ended approach is necessary to determine the scope of existing variables, conditions, or practices that may affect the general goal of the FSR project. The description of indigenous technologies for the management of plants and animals by Michie and Curry, and the analysis of marketing channels by Reeves to understand price differentials which may affect technology use are examples presented in this volume. Nevertheless, there are many more potential outcomes of characterizations. The breadth of potential topics imposes a crucial responsibility on social scientists involved in FSR; in a certain sense, they must be "their own bosses" for delineating relevant areas for investigation. The professionals who carry out a characterization are the first in a project to be aware of, and the best informed regarding local conditions. Decisions must be made as to how to direct the characterization activities, as well as directions for work of other team members.

This phase of work may be best viewed as the generation and testing of hypotheses regarding development conditions in the project area. As preliminary results emerge, they should be presented to other team members or national collaborators for evaluation. Through the interaction of team members, technical and implementation implications for hypotheses will be roughed out in discussions, in an ongoing process. Team members propose alternative hypotheses to be confirmed in the field until a consensus is reached regarding the most appropriate interpretation of conditions.

The increasing responsibility of social scientists within agricultural development projects is nowhere better illustrated than when they direct projects as in the cases of Robins and Hansen. When social scientists are not permanent project members, they exercise a rather free-ranging "gadfly" role, free to point out errors or debatable decisions made, but with a minimal responsibility for overall project development. Unfortunately, this sort of professional relationship is not constructive because it does not permit a careful analysis of problems and alternative solutions. It is not uncommon that questionable decisions are made in the face of problematic social conditions, not due to an unawareness of the problem, but for a lack of knowledge of viable alternatives. As project directors, social scientists have greater authority to direct project activities in directions they find appropriate, but at the same time are put in the position of making difficult decisions not only of a technical nature but of a politico-administrative nature as well.

Another area of FSR methodology where social scientists have been active is in the question of technology evaluation and adaptation. At one time it was believed that questions of technology were straightforward; either a technology worked, or it didn't. It is now clear that technologies usually work in certain contexts, or that certain aspects of a technology will be useful and others less so. Analyses of technology may be restricted to an evaluation of applicability (as in Jones' analysis), or may be directed toward the design of parameters for useful technologies (as in Chapman), or the design of validation and demonstration trials (Hansen, and Wallace and Ahsan). These analyses are typical "normal science" activities to determine whether elements in the proposed technology interact with the farming system as predicted.

The involvement of social scientists in the design and testing of new technologies is linked to the "post-revolutionary" phase in which agricultural development now finds itself. There are no revolutionary new techniques or inputs which are likely to gain universal acceptance. New adaptations of the existing technology face a maze of constraining factors due to unrecognized biological factors or interactions which arise between system components, and due to socioeconomic limitations such as prices (either on the input or the output side), or social characteristics of the group of beneficiaries toward which the technology is directed. The need for social science input in technology generation, first noted by Hildebrand (1976), has been confirmed by later experience in FSR.

In FSR, the research process is closely linked to the process of implementation. Just as technology evaluation and validation can be viewed as a limited scale form of extension, extension can be conversely viewed as a large scale validation. Rhoades, Horton and Booth, and Brush discuss how closely the direction of team research are tied to implementation activities. Nevertheless, as Spring's work demonstrates, questions more strictly concerned with extension methodology will inevitably arise as target populations are identified and implementation problems are detected.

Nowhere is the post-revolutionary condition of agricultural technology more evident than in the need to work beyond the confines of the research station. Farmer adaptations of improved crop varieties and the use of locally developed crop varieties create regional cropping systems which may be improved through scientific research. In both cases, the investigation of off-station techniques can be equally or more important than on-station investigation in terms of short-term payoff and long-term acceptability.

The most positive outcome of the involvement of social scientists in FSR teams is the opportunity to develop more incisive problem foci and methodologies through interaction with other project members. Many biological researchers, or project directors trained in agricultural sciences, are concerned about "social" conditions but are not comfortable with social science methodologies. Misunderstandings arise at times due to a confusion regarding the nature of social phenomena, but in other cases the confusion can arise out of methodological factors, such as the use of certain statistical procedures, and can be resolved to the mutual satisfaction of social and biological researchers (as was done in the fieldwork described by Jones and by Hansen). It is revealing that "social" problems as defined by non-social scientists often include marketing, land tenure, motivation, attitudes, as well as management questions and decision-making criteria. Since these sorts of questions fall so far outside the competence of biological scientists, they are at a loss as to how to begin to describe or categorize them for analysis, and often are greatly relieved to have a resident "expert" to help unravel these problems which they sense but cannot grasp in a methodological sense.

While the experience in FSR has reaffirmed the importance of social scientists as cultural "brokers" between technicians and clients in the process of technology development and transfer, it has also pointed to the need for social scientists to understand existing farmer technologies and scientific research processes. An overly zealous concentration on social factors by social scientists in development makes cooperation difficult with other team members, and may lead as well to the misattribution of causes of problems due to a poor understanding of technical questions involved. Social science is one of several tools used in agricultural development, and the most important rule in its use is to apply it when it is the best tool to resolve the problem at hand.

Continuing Tensions in FSR

A continuing tension exists between the perspectives of individual scientists involved in FSR teams. While the objectives of implementation projects must be clearly stated for a project to begin, individual scientists must respond to specific and often unforeseen problems which arise within the project framework. The problems encountered by projects often have debatable causes and solutions, which may well bring team members into conflict regarding the necessary response by the team.

The need to employ "experts" in the process of agricultural development derives from the extreme variability in agricultural development conditions. Were it possible to prescribe a simple set of procedures to achieve project goals, cook-book plans could be developed at the beginning of each project, and "expert" teams could be replaced by "less-expert" teams whose job would be to implement rather than interpret. Unfortunately, the constant flux of conditions make it difficult to develop unambiguous implementation plans. Experts are expected to constantly adapt their activities to the needs of the situation, according to their knowledge and experience.

The individuality implied in the activities of experts is one of the most divisive tendencies in FSR. Each expert on the team responds to concrete observations and "intuitive" understandings to redesign their work plans. Intuitive understandings are those which the expert has not been able to codify within an explicit logical framework, but which are strongly felt to be true. The continued functioning of an interdisciplinary team depends on the ability to formalize intuitions, and make them accessible to other team members, so an agreement can be reached about how to proceed, or alternatively, a set of testable hypotheses developed to resolve these questions. Individuality and intuition are special skills which experts bring to projects, but these same skills have destructive potential within implementation teams. The two case studies from CIP discuss the problem very candidly, and demonstrate the time and emotional energy devoted to "battling-it-out" to a consensus within the team. In the best of cases, individual team members find their convictions vindicated by later developments, and by turns, discover a new respect for other team members and new forms of cooperation. In the worst cases, institutional arrangements may not demand the continuing of the debate, and interdisciplinary teams may degenerate into multidisciplinary groups with minimal constructive interaction.

The tension between "applied" and "basic" research can be one manifestation of the problems mentioned above. A "basic" research problem may be directed toward determining the cause of an immediate problem, rather than resolving the problem itself. The proper focus is clearly debatable in such cases, but the discussion should be of great importance to all team members. Continuing such a debate within the project framework has the advantage that solutions can be implemented as part of the project, and not be viewed as external criticism of project execution.

An even stronger case can be made for continuing debates regarding "basic" and "applied" research from a pragmatic social science point of view. A major limitation in social science investigation is that its relevance is not understood by potential users of that information. It is not uncommon for agricultural development efforts to spin off long-term biological investigation projects on the basis of problems discovered in the implementation process, but it is much less common with social science problems. The definition of research problems through team interactions helps social scientists to identify investigations with a broad applicability, and at the same time helps non-social scientists to understand why these research problems are important for their own activities.

The Future of FSR and Lessons for Social Science

FSR has enjoyed a period of fashionability which spawned projects all over the world. Various international development funding agencies have decided to focus on FSR as a method for achieving their development objectives, but high expectations may prove to be the demise of FSR. The methodology has had near miraculous powers attributed to it which all but guarantees that donors and clients will be disappointed (Gilbert, Norman and Winch 1980; DeWalt 1985), since there are certainly situations where it is not the most appropriate approach.

The intense interest in FSR has generated a fund of "easy money" which in some cases has been used to promote activities which are not directly related to FSR. The flexibility of the methodology, and the growing number of projects increase the probability that there will be some mistakes made which will most likely be attributed to failures in the methodology, rather than problems in the funding process.

One reaction to the exuberant expansion of FSR projects and practicioners has been a certain reserve on the part of the earlier proponents of the methodology. Several international, regional and national research and implementation institutions have begun to distance themselves from "FSR" to avoid having their priorities or procedures associated with, or dictated by, new FSR developments. This reaction may be entirely consistent with the concept of FSR, which was developed to recognize the specificity of each production situation, in an agronomic and social sense. The "universalization" of FSR is to some extent a threat to the freedom to move into new problem areas as they are detected.

While the future for "FSR" as a named entity may be gloomy, there are undeniable advances which the methodology has contributed to international development, and which will most likely persist in some form. Many of these advances have to do with the oft-mentioned "holism" of FSR, and the interest in defining the full range of biological and non-biological factors which may affect technology adoption or use.

The increased appreciation for indigenous technical knowledge is a major breakthrough which has come to be associated with FSR. The use of on-farm trials has invited direct comparisons of "improved" technologies with traditional technologies, but on the traditional technologies' terms. Erratic environmental conditions, unpredictable markets, interaction problems with other farm components, etc., have been closely observed in the management of on-farm trials and validations, and the advantages of traditional technologies have been better understood thereby.

The development of the interdisciplinary team concept has been an important contribution to social science methodologies. Prior research strategies involving social science either required that the social scientist be given a limited task with regard to demographic or crop distribution questions, or be left to do whatever he or she independently defined as relevant. There was little call for interaction in the definition of research problems or methodologies, and it was to a large extent "hit-or-miss" with regard to the production of useful project information, with only a few notable exceptions. The participation of social scientists on interdisciplinary teams has led to a better appreciation of implementation constraints, especially in terms of time and resources available for information collection. "Hybrid" social science methodologies have developed to respond to the needs of projects. At the same time, there has been an expansion in the numbers of a previously rare professional, the social scientist able to communicate social science concerns to non-social scientists, and capable of responding to technical problems as they arise.

There has also been an increased appreciation for the "social" content of agricultural research activities. On-farm trials have emphasized the need to adapt research station technologies to local socioeconomic conditions. Whereas the role of the social scientist was formerly seen to be one of convincing farmers how to use new technologies, there is a new awareness that lines of communication must be developed to understand farmer dissatisfaction with new technology. On-station research must direct itself toward the resolution of the problems which in fact limit farmer productivity. These problems are not always solved by higher yielding varieties, and may require an understanding of local production conditions or internal family dynamics as a prerequisite to developing acceptable technologies.

During the first half of the twentieth century, world agriculture lived a revolutionary euphoria. A series of techniques which dramatically improved production were discovered and diffused, first in the "developed" countries, and later in the "less-developed" countries. It may even be argued that the "Green Revolution" was the tail-end of that period, in that international genetic research centers permitted the diffusion of hybrid seeds and the massive use of fertilizers to new regions of the world. In the aftermath of these earlier changes, agriculture seems to have stagnated; agricultural innovations no longer meet with the near universal applicability of some earlier innovations. Agricultural development has moved into a phase requiring the painstaking implementation of refinements in technologies. Improvements are harder to come by, and yield improvements are much less dramatic; increased income must be coaxed out of these technologies within established social and technical paradigms. The methodologies and work patterns developed in FSR have provided a basis for the increased participation of social scientists in the painstaking work of cobbling together new technologies which will improve the lives of producers.

Future Needs for FSR

This review of agricultural implementation projects suggests one broad pattern in current activities, which is an emphasis on the introduction of management improvements or adjustments to existing production systems, rather than crop introductions or the promotion of dramatically different new techniques or inputs. These projects are characterized by a careful, sometimes extensive initial evaluation of production conditions, and the recommendation of adjustments in factor mixes, or crop mixes to achieve more efficient production. With the diffusion of agricultural research centers, it is likely to become more common that farmers will have encountered "new" technologies before they are introduced through extension programs. Field trials, commercial interests, technicians' own farms, and wealthy, educated or innovative farmers, all are potential points of diffusion of new technologies. Technical improvement and extension programs must direct themselves to the level of technical advance in their project area to avoid the rueful situation encountered by Hansen, where farmers had already advanced beyond the agricultural technicians in technical and implementation knowledge.

Experience in Central America demonstrates this pattern of technical change. The use of **Pennisetum** spp. and **Brachiaria** spp. forages in Costa Rica and Panama spreads from seed material acquired from field tests of these species by research institutions; farmers have been known to steal seed (plant material, in this case) in small quantities to begin their own seed beds long before the tests reach their stage of "recommendation". In Honduras **Leucaena** spp. was introduced for forage and fuelwood, but while trials were still being carried out by technicians in 10m by 10m plots, local peasants had collected seed from the trials, and planted trees by the thousands on the basis of their observations of the initial results. CATIE technicians familiar with the experiment only half jokingly observed that by the time the final results from the field trials were evaluated, the entire valley would have been planted in **Leucaena** by local farmers.

The fact that many agricultural innovations have now spread "under their own power", so to speak, should not be taken to mean that agricultural development has now been accomplished. While yields have been improved in many cases, there still remain important possibilities for improvement. More importantly, there exists a need to counteract hazardous tendencies which have spun off from new technologies. One mixed blessing of agricultural development has been the increased capability to farm formerly "non-agricultural" lands. Many lands that were problematic for agricultural production due to low fertility, high rainfall, high costs of clearing, etc., are now cultivable with the use of chemical fertilizers, mechanical land cultivation devices, and herbicides. While it cannot be denied that there are poor farmers who have been given a new opportunity for self sufficiency through the clearing and colonization of new land, this process presents grave dangers in a longer time-frame.

The problems facing agricultural development, and especially FSR in the future are 1) the identification of existing production systems which may be improved through research 2) the fine-tuning of techniques to optimize resource utilization and 3) the promotion of techniques to mitigate some of the side effects which have arisen out of the first decades of development.

In the course of their production activities, farmers often create or adapt production systems to special problems of the local environment, but the trend toward the increasing dependence on agrochemicals has eclipsed techniques which are potentially more economical and which eliminate undesirable side effects encountered in chemical use. For example, farmers have developed many management strategies which involve nitrogen-fixing plants. Since nitrogen is a major component of purchased inputs for improving both plant and animal production, the use of nitrogen fixing plants in the tropics is a potential source of savings on agrochemical purchases, as well as a contribution to improved land management. Nitrogen fixing plants, and especially trees, are uniquely adapted to tropical conditions. Deep root systems protect against the problems of periodic drought and erosion experienced in many areas, and nitrogen fixation permits the utilization of lands otherwise undesirable due to low soil fertility. Leaf litter is a fertilizer and a soil protector. The leaves of nitrogen fixing plants often have high protein contents, offering possibilities for producing high protein animal feeds for milk or small animal production, and thereby reorienting the current attempt to supply protein through extensive grazing practices. Some nitrogen fixing tree species produce exceptionally valuable woods (Rosewood, for example, comes from the nitrogen fixing **Dalbergia** species) while others produce good hardwoods and are known to be fast growing species, such as **Leucaena** and **Albizia** (NAS 1979). The tropics are home to many native nitrogen fixing plants which have a great potential contribution to development goals, although this requires the readaptation of new chemical techniques. The use of selective herbicides is a widely used pasture management tool that leads to the elimination of naturally occurring nitrogen fixing plants in pastures; an ironic situation is developing where nitrogen fixing pasture seed is being imported from other countries to replace native nitrogen fixing species which are eliminated through herbicide use! Agricultural development schemes must move in the direction of development plans which combine broad considerations of long term resource management, short term farm management and production improvement. Nitrogen-fixing plants offer a variety of uses which contribute to these goals (Budowski 1985).

The fine-tuning of spontaneously diffused agricultural technologies presents an important area where FSR can help improve on local practices. FSR's characterization, technology design and validation stages attempt to do this, but inevitably the activities are highly directed toward project objectives, and must be carried out in restrictive time frames. In a certain sense, there was a time when it was sufficient to describe a local agriculture as "traditional", for this implied that a standard package of hybrid seeds and agrochemicals could improve production. This distinction has become much less clear, and many "traditional" agriculturalists make extensive use of both improved seeds and agrochemicals. Detailed ethnographic descriptions of "semi-improved" production systems will be important for the design of new and improved systems, although these sorts of descriptions are not normally contemplated within the FSR framework.

It is not uncommon to find misuse of new technologies, either through farmers' misconceptions or at the prompting of agrochemical salespeople. Overdosages or badly timed applications of chemicals raise costs of production and decrease farmer income. The overdosage of agrochemicals is a manifestation of agricultural risk aversion; farmers comment that the cost of additional chemical applications is preferable to the risk of losing a crop. Since most agrochemicals are imported, these misapplications also have a negative effect on national economies, in addition to health and environmental conditions. Conversely, the use of low dosages of chemical controls promotes the development of pests resistant to those controls. The diffusion of many innovations has now taken place, and the task that remains is one of applying techniques to ensure that these innovations will result in a maximum improvement in the farm family income without damaging the quality of life over a longer period.

Another problem with which FSR must begin to grapple is that of the long-term side effects of improved techniques and inputs, specifically deforestation for inappropriate land-use, and environmental contamination. Much of the damage caused by deforestation is removed in space and time from the actual area of deforestation. Even though erosion and flooding are the most commonly cited problems, the farmers themselves also face consequences. Local water sources are damaged by erosion and decreased infiltration within only a few years of deforestation. Farmers also confront a scarcity of fuelwood and building materials, to mention only a few immediate problems (Jones and

Price 1985). But a condition which applies to both locally felt and non-locally felt damage is that the current phase of change in land use is occurring at a rate much more rapid than previously experienced. In Central America, the colonization of the most desirable lands lasted nearly 500 years; in contrast, the occupation of the least desirable lands (which are more than half of the land area) has occurred in the past 50 years. Population change in upper and lower watersheds has accompanied an equally rapid change in hydrologic regimes, so patterns of flooding and water availability upon which settlements are planned are quickly modified. Populations caught in this trap have neither technical nor legislative mechanisms to deal with the conditions which present themselves (Jones in press).

One challenge for agricultural development is the design of systems which are stable within these new environments. The introduction of permanent cropping regimes, minimum tillage for annual cropping, erosion control measures, and agroforestry systems are measures which might be adapted to these new environments. In many areas farmers have already begun to experiment with similar systems, but in others agricultural researchers will have to make these introductions. The rapid changes in agricultural technology in the past have opened agricultural areas which are inherently unstable, and development activities must attend both the farmers' short term production problems and the countries' longer term problem of sustaining or improving levels of production without suffering degradation in the future. While this broad regional perspective appears in some FSR frameworks, it is often either ignored or subordinated to more immediate concerns (such as production).

Environmental contamination is of increasing concern in developing countries. Whereas this problem used to be associated with industrial areas, agriculture now is a major source of contaminants. In Central America, the major limitation to colonization was the existence of a dry season which would permit brush to be burned off. In areas of high rainfall, weed growth is so exuberant that maintaining agricultural lands without the use of fire is extremely costly. This changed with the introduction of herbicides which could work with a minimal dry period for their application. Even in areas where fire was previously used, herbicides have been introduced as technical "improvements". Millions of hectares in Central America alone are managed with periodic applications of herbicides. Similarly, the introduction of fungicides has been at once a technological advance and an environmental and health hazard. In areas of high rainfall and humidity, fungus attacks are a major cause of crop losses. The introduction of fungicides has permitted the cultivation of vegetable crops which were previously impossible, but which require massive applications of fungicides to keep diseases under control. The widely distributed use of chemicals, and minimal capacities of oversight agencies for control over misuse (especially during later stages of cultivation when chemical use should be discontinued to "detoxify" plants) present potentially serious health hazards to human populations.

A major task which currently faces FSR, and agricultural development in general, is the generation of new production strategies which are appropriate to tropical environments. The major areas for agricultural expansion are in humid tropical areas which had been relatively ignored due to their exceptional production problems. Intensive grain production techniques involving annual crops have been transferred on a large scale into these humid environments, but with the exception of paddy rice, this strategy suffers a comparative disadvantage **vis-a-vis** drier areas; the maintenance of soil fertility, and the control of weed invasions, insects, diseases, and erosion all are more costly in humid areas. Animal production must contend with parasites and diseases characteristic of humid areas, as well as the low quality of pastures commonly found. While advances in agricultural technology have permitted the incorporation of new lands into the agricultural landscape in the recent past, the development process is quickly running up against limits within these new environments. Problems are arising not only with the use of specific chemical inputs and genetic material, but in the management strategy and legislative and legal framework for land use which have been generally oriented toward the problems of more "traditional" environments in temperate parts of the world. Major changes are required to minimize pressures for continuing misuse of new production environments now that new technologies permit their more intensive exploitation. This is indeed a tall order for FSR, but it is a necessary achievement to avoid that the implementation of new technologies does not contribute to an accelerated degradation of resources and quality of life as inappropriately managed resource bases degenerate.

The greatest challenge, and possibly the determining factor in the success or failure of agricultural development is the question of social equity. As can be witnessed from the experience in California, the technical advance of agriculture gives a disproportionate leverage over technical development to a minority of wealthy farmers if not properly managed. It may even lead to the dispossession of farmers by financial management entities when capital intensification of agriculture finds favorable markets and institutional conditions (Goldschmidt

1978). Since technology development is an expensive, and publicly subsidized activity, there must be controls to ensure that outputs are beneficial to the society as a whole, and not a select group. Agricultural development which emphasizes production over social concerns may tend to exacerbate existing social problems, and accelerate changes to the point that existing social compensation mechanisms are insufficient to prevent breakdowns in the economic and social order. A superficial reading of the cases presented here may find that explicitly "social" questions are not discussed at much length. However, the development of technology with clearly defined client populations in mind, and by implication with a clearly defined social outcome, is an underlying concern which runs through FSR research as presented in this volume, and its unspoken acceptance is a measure of the success of pioneering social scientists. Social equity will be the product of technology generation and transfer which is directed toward disadvantaged populations, whose acceptance gives a comparative advantage to those farmers.

In the past, there has been a strong focus on production improvement rather than long term productivity concerns in questions of agricultural development, and advances in agricultural biological and chemical technology now provide development efforts with a large and varied tool kit to resolve producers' problems. Effort must now be directed toward the improved management of these new technologies for the achievement of social, economic and long-term development goals. While the results of such activities will not have the flair that the discovery of new crop varieties does, they are necessary to complete the work of agricultural development begun through research into production improvement. The methodologies developed within the framework of FSR projects at present are the most powerful tool available for completing this work.

Bibliography

Boserup, Ester. 1973.
Las Condiciones del Desarrollo en la Agricultura. Madrid: Editorial Tecnos.

Budowski, Gerardo. 1985.
La Conservacion como Instrumento para el Desarrollo. San Jose, Costa Rica: Editorial

Universidad Estatal a Distancia.

Cancian, Frank. 1965.
Economics and Prestige in a Maya Community. Stanford: Stanford University Press.

Carter, William. 1969.
New Land and Old Traditions. Gainesville: University of Florida Press.

DeWalt, Billie R. 1985.
"Anthropology, Sociology and Farming Systems Research". **Human Organization** Vol. 44 (2); 106-114.

Espinosa, Jaime. 1985.
"Agroquimicos: Para Que?". In S. Heckadon and J. Espinosa (eds)., **Agonia de la Naturaleza.** Panama: IDIAP and STRI.

Gilbert, Elon H., David W. Norman and Fred E. Winch. 1980.
Farming Systems Research. East Lansing, Michigan: Department of Agricultural Economics, Michigan State University.

Goldschmidt, Walter R. 1978.
As You Sow: Three Studies on the Social Consequences of Agribusiness. Montclair, New Jersey: Allanheld, Osmun and Co.

Heckadon Moreno, Stanley and Jaime Espinosa Gonzalez. 1985.
Agonia de la Naturaleza. Panama: Instituto de Investigacion Agropecuaria de Panama and Smithsonian Tropical Research Center.

Hildebrand, Peter E. 1976.
"Generating Technology for Traditional Farmers: A Multidisciplinary Methodology. "Paper prepared for the conference on" Developing Economies in Agrarian Regions". Bellagio, Italy.

Jones, Jeffrey R. 1976.
"Market Imperfections in Peasant Societies". In Loucky, J. and J. Jones (eds)., **Paths to the Symbolic Self.** Los Angeles: Department of Anthropology, UCLA.

———. In Press.
Land Colonization in Central America. Tokyo, Japan: United Nations University.

Jones, Jeffrey R. and Norman Price. 1985.
"Agroforestry: An Application of the Farming Systems Research to Forestry". **Human Organization** 44(4); 322 - 331.

Kuhn, Thomas. 1962.
The Structure of Scientific Revolutions. Chicago: University of Chicago Press.

NAS (National Academy of Sciences). 1979.
Tropical Legumes: Resources for the Future. Washington, DC. National Academy of Sciences.

Rappaport, Roy. 1968.
Pigs for the Ancestors. New Haven: Yale University Press.

Whyte, William F. and Damon Boynton. 1985.
Higher-Yielding Human Systems for Agriculture. Ithaca: Cornell University Press.

Biographies of Contributors

Ekramul Ahsan is Member-Director of Agricultural Economics and Social Sciences, and Chairman of the Bangladesh Agricultural Research Council. He holds a Ph.D. in Economics from the University of Hawaii. His primary research interest is in fertilizer use improvement, and he is also interested in techniques of institutional networking for agricultural development.

Robert Booth is British and obtained both his B.Sc. in Botany (1965) and his Ph.D. in plant pathology (1968) from the University of Manchester, England. After a short period of employment at the National Institute of Agricultural Botany in Cambridge, England, he joined the Tropical Products Institute (TPI) of the British Ministry for Overseas Development where he specialized in research on the post-harvest behavior and storage of tropical root crops. For the TPI he undertook numerous short- and long-term overseas technical assistance assignments including three years at the Centro Internacional de Agricultura Tropical in Colombia and two years at the Malaysian Agricultural Research and Development Institute in Malaysia. Since 1978 he has been a member and more recently the leader of the post-harvest technology team at CIP.

Stephen B. Brush is Professor in the International Agricultural Development Program, at the University of California, Davis. He formerly taught at the College of William and Mary, and has served on the review board for the National Science Foundation.

James Chapman did his dissertation research at the International Rice Research Institute at Los Banos, Philippines. He has worked at the Interamerican Institute for Agricultural Cooperation (IICA) in Coronado, Costa Rica, as agricultural economist, and is presently directing a project for improving irrigation in Dominican Republic.

John J. Curry, Jr. is a member of the Swaziland Cropping Systems and Extension Training Program.

Peter E. Hildebrand is Professor in the Food and Resource Economics Department at the University of Florida. Previously, he was with the Rockefeller Foundation and on the faculties of the University of Nebraska, Colorado State University, and Texas A&M University. He has been involved in development projects since 1964 in Pakistan, Colombia, El Salvador, and Guatemala.

Art Hansen is Associate Professor and Graduate Coordinator, Department of Anthropology, University of Florida, and the Chair of the Faculty Steering Committee of the Food in Africa Program of the Center for African Studies. He has 9 years of international research and development experience in Bolivia, the Dominican Republic, Zambia, and Malawi and has co-edited 3 books, the most recent being **Food in Sub-Saharan Africa** (Lynne Rienner Publishers 1986).

Douglas E. Horton, a native of midwestern USA has his B.S. and M.S. degrees in farm management and agricultural economics from the University of Illinois and a Ph.D. in economics from Cornell University. After brief periods of employment at the University of Wisconsin and the World Bank, he has worked at CIP since 1975.

Jeffrey R. Jones is Assistant Professor in the International Development Studies Program at Clark University. He previously worked as staff social scientist at CATIE, in Costa Rica, and has worked in agricultural development and natural resource implementation projects in Guatemala, Honduras, Nicaragua, Costa Rica, and Panama. Major research interests are tropical land use, land settlement and community forestry. He is currently involved in research on the Indigenous Technical Knowledge of nitrogen-fixing trees and the evaluation of experiences in tropical land colonization in Central America.

Barry H. Michie is an economic anthropologist who works with agrarian and developmental issues. At Kansas State University since 1976 he has been active in the establishment of an on-campus FSR program, as Assistant Director of the Agriculture and Liberal Arts Program, and served as principal investigator of an FSR project in India. His major research has been conducted in India with special emphasis on changing structures of agrarian economy, traditional agricultural production systems, and applied agricultural research.

Edward B. Reeves has a Ph.D. in Anthropology from the University of Kentucky and is an assistant professor of sociology and anthropology at Morehead State University. Dr. Reeves has done extensive field research in Egypt as well as the Western Sudan. He has been involved in Farming Systems Research and Development since 1981 and currently is making a study of limited-resource farmers in Appalachia.

Robert E. Rhoades, a citizen of the USA, holds a B.S. in agriculture, M.S. in sociology, and Ph.D. in anthropology. He began his development career as an agricultural extension worker in Asia (1962-64) only to discover years later the value of social science input into agricultural research and development. He has worked at the International Potato Center (CIP) in Peru since 1979.

Edward Robins teaches in the Department of Sociology, University of Wisconsin, River Falls, and serves as social science advisor to the AID mission in Rwanda, under the Joint Career Corps Program.

Anita Spring is Associate Professor in the Department of Anthropology and Associate Dean of the College of Liberal Arts and Sciences at the University of Florida where she also directs the Women in Agriculture Program. She has carried out research in Zambia, Malawi, and Cameroon on agricultural development, health care systems and gender issues. She is co-editor of **Women in Ritual and Symbolic Roles** (1978) and **Women Creating Wealth: Transforming Economic Development** (1985) and author of **Agricultural Development in Malawi: A Project for Women in Development.**

Ben J. Wallace is Professor of Anthropology, Southern Methodist University and Consultant to International Agricultural Development Services (Bangladesh Project) and the Bangladesh Agricultural Research Council. He has worked twenty years in research and teaching in the socio-cultural dimensions of Farming Systems, especially in Southeast Asia and South Asia. He is author of numerous articles and books, and recently has published the book **Working the Land: A Case Study in Applied Anthropology and Farming Systems Research in Bangladesh.** His current research analyzes the contribution of females to the rural economy of Bangladesh.